The Place Where You Go to Listen

JOHN LUTHER ADAMS

The Place Where You Go to Listen

IN SEARCH OF
AN ECOLOGY OF
MUSIC

WESLEYAN UNIVERSITY PRESS MIDDLETOWN, CONNECTICUT

Published by Wesleyan University Press,
Middletown, CT 06459
www.wesleyan.edu/wespress
© 2009 by John Luther Adams

Foreword by Alex Ross excerpted from
"Song of the Earth," *The New Yorker,* May 12, 2008.
Used by permission.

Wesleyan University Press is a member of
the GreenPress Initiative. The paper used in
this book meets their minimum requirement
for recycled paper.

Library of Congress
Cataloging-in-Publication Data
Adams, John Luther, 1953–
The place where you go to listen : in search of an
ecology of music / John Luther Adams ; foreword
by Alex Ross.
    p. cm.
Includes bibliographical references.
ISBN 978-0-8195-6903-5 (pbk.)
1. Adams, John Luther, 1953– 2. Composers—
Alaska—Biography. 3. Adams, John Luther,
1953– Place where you go to listen. I. Title.
ML410.A2333A3 2009
780.92—dc22
[B]   2008054108

Contents

Illustrations

Foreword ALEX ROSS

On a recent trip to the Alaskan interior, I didn't get to see the aurora borealis, but I did, in a way, hear it. At the Museum of the North, on the grounds of the University of Alaska in Fairbanks, the composer John Luther Adams has created a sound-and-light installation called *The Place Where You Go to Listen*—a kind of infinite musical work that is controlled by natural events occurring in real time. The title refers to Naalagiagvik, a place on the coast of the Arctic Ocean where, according to legend, a spiritually attuned Iñupiaq woman went to hear the voices of birds, whales and unseen things around her. In keeping with that magical idea, the mechanism of *The Place* translates raw data into music: information from seismological, meteorological and geomagnetic stations in various parts of Alaska is fed into a computer and transformed into an intricate, vibrantly colored field of electronic sound.

The Place occupies a small white-walled room on the museum's second floor. You sit on a bench before five glass panels, which change color according to the time of day and the season. What you notice first is a dense, organlike sonority, which Adams has named the Day Choir. Its notes follow the contour of the natural harmonic series—the rainbow of overtones that emanate from a vibrating string—and have the brightness of music in a major key. In overcast weather, the harmonies are relatively narrow in range; when the sun comes out, they stretch across four octaves. After the sun goes down, a darker, moodier set of chords, the Night Choir, moves to the forefront. The moon is audible as a narrow sliver of noise. Pulsating patterns in the bass, which Adams calls Earth Drums, are activated by small earthquakes and other seismic events around Alaska. And shimmering sounds in the extreme registers—the Aurora Bells—are tied to the fluctuations in the magnetic field that cause the northern lights.

The first day I was there, *The Place* was subdued, though it cast a hypnotic spell. Checking the Alaskan data stations on my laptop, I saw that geomagnetic activity was negligible. Some minor seismic activity in the region had set off the bass frequencies, but it was a rather opaque ripple of beats, suggestive of a dance party in an underground crypt. Clouds covered the sky, so the Day Choir was muted. After a few minutes, there was a noticeable change: the solar harmonies acquired extra radiance, with upper intervals oscillating in an almost melodic fashion. Certain that the sun had come out, I left *The Place* and looked out the windows of the lobby. The Alaska Range was glistening on the far side of the Tanana Valley.

When I arrived the next day, just before noon, *The Place* was jumping. A mild earthquake in the Alaska Range, measuring 2.99 on the Richter scale, was causing the Earth Drums to pound more loudly and go deeper in register. (If a major earthquake were to hit Fairbanks, *The Place,* if it survived, would throb to the frequency 24.27 Hz, an abyssal tone that Adams associates with the rotation of the earth.) Even more spectacular were the high sounds showering down from speakers on the ceiling. On the Web site of the University of Alaska's Geophysical Institute, aurora activity was rated 5 on a scale from 0 to 9, or "active." This was sufficient to make the Aurora Bells come alive. The Day and Night Choirs follow the equal-tempered tuning used by most Western instruments, but the Bells are filtered through a different harmonic prism, one determined by various series of prime numbers. I had the impression of a carillon ringing miles above the earth.

On the two days I visited *The Place,* various tourists came and went. Some, armed with cameras and guidebooks, stood against the back wall, looking alarmed, and left quickly. Others were entranced. One young woman assumed a yoga position and meditated; she took *The Place* to be a specimen of ambient music, the kind of thing you can bliss out to, and she wasn't entirely mistaken. At the same time, it is a forbiddingly complex creation that contains a probably irresolvable philosophical contradiction. On the one hand, it lacks a will of its own; it is at the mercy of its data streams, the humors of the earth. On the other hand, it is a deeply personal work, whose material reflects Adams's long-standing preoccupation with multiple systems of tuning, his fascination with slow-motion formal processes, his love of foggy masses of sound in which many events unfold at independent tempos.

The Place, which opened on the spring equinox in 2006, confirms Adams's status as one of the most original musical thinkers of the new century. At the age of fifty-five, he is perhaps the chief standard-bearer of American experimental music, of the tradition of solitary sonic tinkering that began on the West Coast almost a century ago and gained new strength after the Second World War, when John Cage and Morton Feldman created supreme abstractions in musical form. Talking about his work, Adams admits that it can sound strange, that it lacks familiar reference points, that it's not exactly popular—by a twist of fate, he is sometimes confused with John Coolidge Adams, the creator of the opera *Nixon in China* and the most widely performed of living American composers—and yet he'll also say that it's got something or, at least, "It's not nothing."

Above all, Adams strives to create musical counterparts to the geography, ecology and native culture of his home state, where he has lived since 1978. He does this not merely by giving his compositions evocative titles—his catalog includes *Earth and the Great Weather, In the White Silence, Strange and Sacred Noise, Dark Waves*—but by literally anchoring the work in the landscapes that have inspired it.

"My music is going inexorably from being about place to *becoming* place," Adams said of his installation.

I have a vivid memory of flying out of Alaska early one morning on my way to Oberlin, where I taught for a couple of fall semesters. It was a glorious early-fall day. Winter was coming in. I love winter, and I didn't want to go. As we crested the central peaks of the Alaska Range, I looked down at Mt. Hayes, and all at once I was overcome by the intense love that I have for this place—an almost erotic feeling about those mountains. Over the next fifteen minutes, I found myself furiously sketching, and when I came up for air I realized, There it is. I knew that I wanted to hear the unheard, that I wanted to somehow transpose the music that is just beyond the reach of our ears into audible vibrations. I knew that it had to be its own space. And I knew that it had to be real—that I couldn't fake this, that nothing could be recorded. It had to have the ring of truth.

Actually, my original conception for "The Place" was truly grandiose. I thought that it might be a piece that could be realized at any location on the earth, and that each location would have its unique sonic signature. That idea—tuning the whole world—stayed with me for a long time. But at some point I realized that I was tuning it so that *this* place, this room, on this hill, looking out over the Alaska Range, was the sweetest-sounding spot on earth.

If life is a journey, then a life's work is a path we follow throughout the journey. My path is music. The music always extends ahead of me, leading me onward. I do my best to follow. Sometimes the path leads me beyond the limit of my understanding, and I find myself lost in what seems to be the middle of nowhere. At times like these I feel most deeply the wonder of the journey.

This book comes from such a time. This is a narrative of artistic exploration and discovery, of the gradual revelation, the day-by-day unfolding, of my largest, most complex work to date. It's also an account of the work itself, of the forms and sources, the processes and forces that constitute and animate it. And this is a statement of poetics, of the aesthetic and philosophical ideals on which the work is grounded and toward which it leads.

The first vision of *The Place Where You Go to Listen* came to me at thirty thousand feet, as I was flying over the peaks of the Alaska Range. From that altitude this new work appeared as vast and clear as the terrain below. In the years that followed—on the ground, in my studio and inside the space of the work—I came gradually to discover the varied topology, the ever-changing light and weather of this artistic terrain in intimate detail.

Like coming to know a place, discovering a work of art takes time. When I began *The Place* I knew it would occupy me for several years. But I couldn't have imagined the many intricate and interwoven layers it would contain. The process of creating this work has been a great learning for me. And it would have been impossible without the talent, knowledge, generosity and sustained efforts of many people.

It began with Aldona Jonaitis. Aldona believed in *The Place* before I did. When I first described the concept to her, she took a great leap of faith and invited me to create this work in the new building at the Museum of the North. Without Aldona's boldness and determination, *The Place* would not have happened, and this new stage in my creative life might never have begun.

Jim Altieri has been my closest collaborator in this work. First in the studio and then in the museum, Jim has worked with me side by side, day after day. Jim's ears, his intellect, his musicality and his sense of humor are all razor sharp. His extraordinary skill as a programmer took *The Place* out of my mind's ear and gave it audible, visible form in time and in the air. I can't imagine having created it without him.

Curt Szuberla is the creator of SunAngles, the mathematical heart of *The Place*. Curt also created MoonAngles and MoonPhases. If I'd asked him to hang the stars, I have no doubt he'd have been willing and able to do that as well, with infinitesimal accuracy. Curt has taught me more about the ephemeris and epochal time than I ever thought I needed to know. He also makes some pretty fine home brew. I only wish I could have worked in the summer-solstice cymbal crash he requested.

Dirk Lummerzheim is my science guru, the teacher I wish I'd had in school. Dirk has a remarkable gift for explaining complex phenomena in clear and simple terms. And he's endlessly patient, answering half-formed, often repetitive questions. Dirk is passionate about his work, relentlessly open-minded and imaginative. Those late nights watching aurora at Poker Flat and talking with Dirk about science and art are some of my fondest memories from working on *The Place*.

Debi-Lee Wilkinson translates the never-ending dance of the magnetosphere into galaxies of numbers and elegant constellations of graphs. Debi has graciously shared her work to be translated into music, allowing listeners in *The Place* to hear the aurora borealis, even when the sun is up or the sky is cloudy.

Roger Hansen is a longtime rock 'n' roller. He's also the state seismologist for Alaska. When I approached Roger with the idea of rumbling the museum floor with earthquake data, he readily opened all the doors to make this happen. The most crucial of these was the door to the office of Josh Stachnik. Josh is the seismological wizard who solved the far from trivial problem of translating earthquake data into sound in real time. Now whenever the ground moves in Alaska, the reverberations of Josh's work are heard and felt by listeners in *The Place*.

John Olson is the scientific godfather of *The Place*. Early on John helped me zero in on the most promising possibilities for the geophysical phenomena to be heard inside *The*

Place. Our wide-ranging conversations were delightful and inspiring demonstrations of the fact that good scientists are every bit as creative as artists.

At the Museum of the North, Amy Geiger, Dan David, Janet Thompson and Kevin May have welcomed *The Place* into the busy life of the building, while Wanda Chin, Terry Dickey, Jennifer Arseneau and Roger Topp have integrated it into the curatorial and educational programs of the institution. Roger's rich interactive program enhances the experience of visitors to *The Place*, and his crisp graphics grace the pages of this book.

Steve Bouta is tireless in his work to keep the sometimes-temperamental technology functioning. Kerynn Fisher charmed the news media as she introduced them to this strange new work of art. Peggy Hetman has been the guardian angel of *The Place*. Peggy always found a way to make procedural snafus disappear and to keep things moving forward. Chad Stadig was endlessly helpful and unflappable in accommodating artistic considerations within a hardhat zone.

Many experienced tradespeople—painters, electricians, insulators, glaziers, masons and others—brought their skills into *The Place*. At the intersection of art and carpentry, Scott Schuttner, Darrin Rorabaugh, Matt Kasvinsky and Silvan Schuttner made invaluable contributions. Paul Lugin was especially heroic, tackling the nastiest jobs of all with relentless good humor and meticulous attention to detail.

Alexander Nichols was persistent and resourceful in helping me focus my initially vague concept of the lighting design and adapt it to the constraints of the site. Tom Erbe helped get things off the ground in the early stages of programming and sound design. Nathaniel Reichman contributed expert advice on numerous technical details. And I'm grateful to Daniel Curran, Lisa Kauffman and everyone at Genelec for their generous assistance in helping me build the superb sound system for *The Place*.

The support of the Paul Allen Foundation was crucial, providing the time and encouragement I needed to refine the initial vision of *The Place*. Then, with the leadership of Diane Kaplan and Helen Howarth, the Rasmuson Foundation provided the commission and the material resources that allowed this vision to become reality.

As always the love, faith, patience and laughter of my beloved wife, Cynthia, have been the bedrock on which I've worked. It is my greatest joy and blessing to share this life with her.

The Place Where You Go to Listen is one stage of a journey that has led me to many strange and beautiful places. Looking back, I can trace the course of my journey and see how the experience of one place has led me on to another. Looking ahead, I'm filled with gratitude for the feeling that my path is always just beginning.

In Search of an Ecology of Music

The science of ecology is a study of patterns. Ecology examines the intricate patterns that connect organisms and the environments in which they live. Beyond the particular patterns themselves, ecology considers the *totality* of patterns and the larger systems they create.

An ecosystem is a network of patterns, a complex multiplicity of elements that function together as a whole. I conceive of music in a similar way. For me the essence of music is not the specific patterns of harmony, melody, rhythm and timbre. It's the totality of the sound, the larger *wholeness* of the music.

The central truth of ecology is that everything in this world is connected to everything else. The great challenge now facing the human species is to live by this truth. We must reintegrate our fragmented consciousness and learn to live in harmony with the larger patterns of life on earth, or we risk our own extinction.

As a composer, I believe that music can contribute to the awakening of our ecological understanding. By deepening our awareness of our connections to the earth, music can provide a sounding model for the renewal of human consciousness and culture. Over the years this belief has led me from music inspired by the songs of birds, to landscape painting with tones, to elemental noise and beyond, in search of an ecology of music.

The songs of birds first awakened in me a profound longing to feel at home in nature. From this longing grew the vision of a music grounded in deep attention to the natural world, a vision that has been at the heart of my work ever since. In the *songbirdsongs*— a collection of pieces for piccolos and percussion composed from 1974 through 1979— I worked without the aid of recordings. I was determined to discover this music directly from the birds, through firsthand experience listening in the field. In time I found myself listening more and more closely to the music of the field itself.

For more than a decade I composed musical landscapes. My experiences in wild places inspired choral and orchestral works such as *Night Peace* (1977), *A Northern Suite* (1979–

81) and *The Far Country of Sleep* (1988). Beginning with *Dream In White On White* (1992), my music became less pictorial as I aspired to evoke the experience, the *feeling* of being in a place, without direct reference to a particular landscape.

In *Earth and the Great Weather* (1990–93) I combined the Native languages of the Iñupiat and Gwich'in peoples with drumming inspired by their dance rhythms, with music for strings inspired by an Aeolian harp and with my own recordings of natural sounds, to create a "sonic geography" of a sacred place—the Arctic National Wildlife Refuge.

The light in northern latitudes embodies colors and feelings that I've experienced in no other place. After living in Alaska for many years, I came to wonder whether I could somehow convey these colors and feelings in music. Beginning with *The Light That Fills the World* (1999/2001), in a series of works including *Dark Wind* (2001), *The Farthest Place* (2001–2), *The Immeasurable Space of Tones* (2002) and *Red Arc/Blue Veil* (2002), I pursued a music composed entirely of floating fields of color. Yet even as my music became more abstract, it continued to be haunted by landscape.

In a trilogy of concert-length orchestral works, *Clouds of Forgetting, Clouds of Unknowing* (1991–95), *In the White Silence* (1998) and *for Lou Harrison* (2003–4), I aspired to evoke those intimations of the sublime that we sometimes feel in a beautiful landscape. But just as we can find transcendent peace in the beauty of nature, we can also discover a different kind of transcendence in the presence of elemental violence.

Inspired by my encounters with calving glaciers, raging rivers, wildfires and extreme weather, *Strange and Sacred Noise* (1991–97) celebrates noise in the primal forces of nature. In this extended cycle for percussion quartet, I discovered more than I had imagined. Deep within the complex sounds of snare drums, tam-tams, tom-toms, bass drums, bells and air raid sirens, I began to hear voices. These voices had a haunting, almost human quality. And I wanted to hear them alone.

My desire to distill the voices of tone from fields of noise led to *The Mathematics of Resonant Bodies* (2002). I began by composing eight pieces for percussion quartet. The incomparable percussionist Steven Schick recorded these quartets. Then, selectively, I erased them. Through digital processing I removed most of the transient noise from the recordings to reveal the essential tones, the inner voices of the instruments. Into these electronic "auras" I reintroduced the noise of the instruments, composing independent parts for a solo percussionist (once again, Steven Schick). Heard together, the "live" instruments and their recorded auras create complex composite sonorities of noise and tone.

Encouraged by the discoveries of *Mathematics,* I continued my explorations of noise. *Veils* (2004–5) began with pure synthetic noise. This six-hour soundscape is composed from long strands of pink noise rising and falling over the full range of human hearing. Shaping a counterpoint of these lines, I layered them into multiple choirs, each moving at its own tempo. Finally, I passed these veils of noise through a series of "harmonic prisms"— banks of filters tuned to prime-number harmonics (from 11 to 31). The resulting fields of sound fill the air with many tones sounding at any moment. But it's often difficult to distinguish one tone from another. They tend to meld together into rich, ambiguous sonorities in which the higher tones sound like harmonics of the lower tones. The timbres are clear and slightly breathy, like human voices mixed with bowed glass or metal.

The sound of *Veils* saturates physical, tonal and temporal space. But rather than overwhelm the listener, I want to seduce the listener to enter into the sound and remain there for extended periods of time. The melding of rhythm, pitch and timbre creates unified fields of sound. My objective is to leave these fields as untouched as possible, letting them fill time and space with forms and colors as simple and as beautiful as they can be.

My discoveries in the *Veils* led me to embrace synthetic noise at the *prima materia* for my largest work to date, and to contemplate the larger poetics of noise.

THE BREATH OF THE WORLD

Wherever we are, what we hear is mostly noise. When we ignore it,
it disturbs us. When we listen to it, we find it fascinating.
— *John Cage, "The Future of Music: Credo" (1937)*

Most music begins with tone.

In acoustical terms, tone is periodic sound. Tones are produced by regular oscillations. They have relatively simple waveforms.

In perceptual terms, tone is pitched sound. Tones have easily recognizable identities grounded in specific frequencies.

Tones can be precisely tuned. They can be arrayed into scales and chords, and used as building blocks for musical constructions.

Noise is not so easy to control. It is the sound of chaos.

In acoustical terms, noise is aperiodic sound, sound produced by irregular vibrations with complex waveforms.

In perceptual terms, noise is unpitched sound—a diffuse band or field of sound, elusive to focus by the ear.

What, then, is the meaning of music that begins with noise?

In Papua New Guinea, when a Kaluli songmaker searches for a new song, he may camp by a waterfall or a running stream. All the songs in the world are contained in the noise of the water. The songmaker listens carefully, sometimes for days, until he hears the voice of his new song.

Whenever we listen carefully, we come to hear that music is around us all the time. Noise is no longer unwanted sound. It is the breath of the world.

If music grounded in tone is a means of sending messages to the world, then music grounded in noise is a means of receiving messages *from* the world. Noise takes us out of ourselves. It invites communion, leading us to embrace the patterns that connect us to everything around us. As we listen carefully to noise, the whole world becomes music. Rather than a vehicle for self-expression, music becomes a mode of awareness.

After years of composing music grounded in metaphors of space and place, I find that my music has now become more tangibly physical, in a small architectural space that resonates within a larger geographic place. *The Place Where You Go to Listen* (2004–6) is a long-term sound-and-light environment housed in the Museum of the North at the University of Alaska Fairbanks. *The Place* is dedicated to hearing the unheard music of the world around us. The rhythms of sunlight and darkness, the phases of the moon, the seismic vibrations of the earth and the fluctuations of the earth's magnetic field all resonate within this space. Streams of data derived from these geophysical phenomena shape the sound and light of *The Place,* which are synthesized and modulated on a computer, in real time.

LOCUS EX MACHINA

The most important musical instrument of the twentieth century may well have been the microphone. The most important musical instrument of the twenty-first century may prove to be the computer. Just as microphones allow us to hear sounds that aren't readily accessible to the naked ear, we can use computers to transform inaudible forces of nature into audible sound.

The computer is the primary instrument with which *The Place Where You Go to Listen* is created. This new instrument allows me to hear and give voice to visible, tactile, invisible and inaudible vibrations of earth and sky. But the most sophisticated technology is transparent. Although *The Place* is created entirely with electronic technology, the medium is not the message. The primary presence in this space is not the instrument. It is the music of place.

It's perhaps ironic that this imaginary world intended to celebrate our connections to the natural world could not have been created without the machine of the computer. Machines allow us to modify energy, matter and information in ways that would be difficult or impossible without them. Our passion for machines has created the dangerous illusion that we can manipulate the living world in any way we choose. And increasingly powerful machines have given us the ability to wreak destruction on a planetary scale. Yet in spite of their destructive powers, we can also employ machines as creative instruments, to extend the reach of our senses and engage us with the world in new ways.

REAL TIME

For almost two decades now, the computer has been one of the tools I've used in my work. Over the years it's become more and more integral to my process of notating, recording and creating sounding models of my music. But *The Place Where You Go to Listen* is the first music I've made in which all the finished sounds are produced with the computer. This was also my first experience working in so-called *real time*.

I remain skeptical of this term. It seems a bit like double-speak, or an oxymoron. Isn't *all* time equally real or unreal? What exactly do we mean when we say "real time"? I'm still not sure. But I can say without hesitation that this new medium has changed my way of working. It may even have changed my fundamental conception of my work.

Composing a work in conventional musical notation, I usually spend a lot of time in pre-compositional thought. On paper I sketch out instrumentation, forms, harmonies, tempo relationships, rhythms and lines. Then I test my sketches—sometimes at the piano, sometimes on the computer—and rework them on the basis of what I hear. The intellect ratifies and refines the perceptions of the ear. But the final proof is the sound.

Working in real time, I find the ear leading the mind a little more. This new medium offers instant feedback. Of course this is also true for the piano or any other musical instrument. But unlike a single-voice or single-timbre instrument, real-time synthesis presents an orchestra of open-ended possibilities directly to our ears. Instead of playing a sound on one instrument and imagining another, or listening to a recorded sample of the intended instrument, I can hear the actual sound itself. I can modify the sound in a variety of ways and hear it change immediately.

As a result I find myself devoting less of my time to *imagining* and more of my time to *listening*. The more I listen, the better I understand the nature of the sounds and the responses they evoke. As I've always done, I listen and make modifications. Then I listen again. But now time is so accelerated that it almost seems to disappear. In a sense time becomes *less* real. The process of composing becomes more like sculpting, working and reworking a malleable substance of sound, space and time.

While the real-time medium can accelerate the process of composing, it can also slow down the experience of listening. Inside *The Place Where You Go to Listen*, events unfold in the same tempos as in nature. The omnipresent atmosphere of sound and light is shaped by the arcs and rhythms of day and night. The fields of tone and color are always changing. But since things happen in real time, the rate of change is usually too slow to be perceived. Yet over the course of hours, days and months, the changes are increasingly dramatic. From day to night, from winter to summer, *The Place Where You Go to Listen* may look and may sound like two very different places.

The Place also encompasses the distinctive sounds of virtual "drums" and "bells" that resonate with earthquakes and with fluctuations in the magnetosphere. But there can be long periods, hours, even days, between major seismic or geomagnetic events. During such times the sounds associated with these forces are quiet or silent.

Real time is an essential element in the composition and the experience of *The Place Where You Go to Listen*. This is not a predetermined sequence of musical events and lighting scenes. It is a dynamic system of visible and audible forces interacting in a constantly changing environment—a self-contained world connected to and resonating with the real world.

Any time a listener walks through the door, I want the sound and the colors to be beautiful, mysterious and implacable. Like a place in the wilderness, *The Place Where You Go to Listen* requires the visitor to enter into it, to take things on its terms, to pay attention and to find her own way.

WORKING IN PLACE

Simple observation is my most important formal device. . . .
The interrelation of observation, analysis, and memory
become, so to speak, the tools of the trade.
— *Richard Serra*

As a composer I often work far removed from the time and space in which the music is eventually heard. I do much of my work in the solitary space of the studio, creating the imaginary space of the score. Once the score is complete, rehearsal, performance and recording follow in other places. This is the way I've worked for years.

But *The Place Where You Go to Listen* demanded a different way of working. In this new medium there is no master score; there are no tangible instruments and no performing musicians. This has led me to a process grounded in direct observation, listening within and listening *to* the physical acoustical space of the work. As much as a process of composition, my work on *The Place* has been a process of *design*.

The specificity of this work comes from the specificity of the setting in which it is experienced. Creating *The Place* required two years of work in my studio. Yet in a very real sense this work could be completed only in the final listening space itself. Once the room was constructed, it became my studio. Much of the creation of *The Place* could happen only from *inside* the space.

The Place Where You Go to Listen is a nexus between the architectural space in which we listen and the larger geographic place with which the work resonates. *The Place* originates in sunlight and darkness, the electromagnetic weather above us and the movements of the earth beneath our feet. We perceive these phenomena within the visible and audible space of the work. Yet the boundaries of the work exceed the physical boundaries of the space in which we experience it.

The Place resonates sympathetically with the world outside. In turn, I hope it reverberates back into the world. We enter with our everyday perceptions of the world around us. Inside *The Place* we hear and see things differently. When we leave, perhaps we carry some of these new perceptions with us.

The music of *The Place* is produced by natural phenomena. But this is not a scientific demonstration of natural phenomena. It is a work of art. The essence of this work is the sounding of natural forces interacting with the consciousness of the listener. This is not a simulated experience of the natural world. It is a heightened form of experience itself.

The Place resonates with nature. But this is nature filtered through my ears. For the listener I hope this music sounds and feels natural, as though it comes directly from the earth and the sky. Yet the decisions about the timbres, tunings, harmonies and melodic curves, the dynamics, rhythms, counterpoint and musical textures were mine. Despite my desire to remove myself and invite the listener to occupy the central position, *The Place* is still a musical composition. Although I tried to minimize the evidence of my hand, I remain the composer.

TWO MINDS

For me composing is a process of discovering and revealing the fundamental unity of the formal and the sensual, the interior and the exterior. I want my music to embody both the objectivity of form and the subjectivity of the sensual.

Formal structure—the mathematics and geometry of composition—gives the music a sense of objectivity independent of the composer. Sensuality of sound invites the listener to enter into the enveloping presence of the music.

Searching for this unity of sound and form requires the composer to move back and forth, into and out of the music, with two different minds. One mind stands just outside the music, regarding it carefully with a mixture of curiosity and detachment. The other mind stands fully inside the music, immersed in the sheer sensation, the wonder of the sound. To be whole, the music must integrate both these minds.

Art and science also embody two minds, two ways of understanding the world in which we live. Yet these two minds share a more fundamental unity than we sometimes recognize. And they have much to say to one another, especially in our times.

Historically, science has aspired to be objective. Art has been a domain of more subjective experience. Science, we tend to think, examines the external world while art expresses the inner world. But artists and scientists know better. Scientists speak eloquently about the role of intuition and imagination in their work. Artists speak with similar clarity and force about the centrality of observation and analytical thinking in their work.

Both art and science spring from curiosity, which is an innate characteristic of the human mind. Science employs our curiosity to advance our cognitive understanding of the world. Yet reductionism in scientific thought has also led us to regard ourselves as apart from the world, rather than a part *of* the world. This fallacy has led us to dominate life on earth to the extent that our own survival is now threatened.

Art employs curiosity to advance our intuitive understanding of the world. It also fulfills the basic human need to express our feelings. But art grounded exclusively in self-expression can indulge our conceit that we somehow stand above and beyond the rest of life. This pretension only exacerbates our sense of alienation from the earth and other species. Overpopulation, overconsumption, pollution, deforestation and widespread extinction are both symptoms and results of this alienation. Perhaps its ultimate manifestation is human-caused climate change.

The overindulgences of artistic romanticism and the mechanistic worldview of Newtonian science were two sides of the same mind that led us to our present predicament.

Still, art and science can teach us to transcend ourselves, guiding us beyond our anthropocentric obsessions to a more complete and integrated relationship with the earth. Science reminds us of the miracles of the larger world (and the universe beyond) to which we belong. Art reminds us of the essential connections of the spirit that we share with all beings and all things.

Science examines the way things are. Art imagines how things *might* be. Both begin with perception and aspire to achieve understanding. Both science and art search for truth. Whether we regard truth as objective and demonstrable or subjective and provisional, both science and art can lead us toward a broader and deeper understanding of reality. Even as they augment our understanding, science and art heighten our sense of wonder at the strange beauty, astonishing complexity and miraculous unity of creation.

We live in a time of great exploration and discovery. But unlike those of previous eras, the most important explorations of our time are not new places. The most important discoveries are not new phenomena. The great learning of our time is of the endlessly complex and subtle interrelationships between places and organisms, between everything in nature from the subatomic to the cosmic.

With characteristically radical elegance, John Cage defined music as "sounds heard."

The idea that music depends on sound and listening might seem as self-evident as the idea that we're an inseparable part of nature. But both these simple truths challenge us to practice ecological awareness in our individual and our collective lives.

Cage's definition of harmony was "sounds heard together."

Listening to the multiplicity of sounds all around us, we learn to hear the marvelous harmony they create. Hearing this harmony, we come to understand our place within it, how our human voices fit into the larger, endless music of the world.

A Composer's Journal Part I—Always Getting Ready

DECEMBER 21, 2003

It's midwinter, the time of darkness and rebirth. In the deep stillness, I begin again.

The Place Where You Go to Listen is the largest, most complex work I've ever undertaken. It extends my journey of the past three decades toward what I hope will be a rich period of discovery. *The Place* calls me to explore new media and new ways of working. I sense, too, that it may lead me to a new conception of music. After this my work may never be the same.

As I begin this work, I begin this journal. Here I'll record my thoughts and impressions as the work unfolds. This is not a sketchbook for the work itself. This is a chronicle of my thoughts and impressions, my notes from the journey.

DECEMBER 22

As always, it begins with questions: What exactly *is* this work? Is this a musical composition? Or is it an architectural space? Will it be unique to this specific geographic location? Or will it be global in conception? Can it somehow be both?

Will the sounds suggest familiar musical instruments? Will they resemble sounds in nature? How loud, how soft will the sounds be? How continuous? Will there ever be silence?

And what about the lighting? What colors will it encompass? How bright, how muted will they be? Will the space ever be completely dark?

The instruments will be electronic, the sounds and colors entirely synthetic. Yet can this *Place* somehow convey the richness and subtlety of acoustical sound and natural light? Can it sound and feel *natural?*

JANUARY 4, 2004

As I begin my research for *The Place Where You Go to Listen,* I'm working hard to finish another large project. For much of this past year I've been composing *for Lou Harrison,* an hour-long piece for string quartet, two pianos and string orchestra. I was not commissioned to write this new piece. I was compelled to do so, in response to the passing of one of the most important people in my life. Lou was a role model, a wise mentor and a generous friend to me for thirty years. I hope this new work expresses something of the love and gratitude I feel toward Lou, perhaps echoing a little of the exuberance of his life and music.

Stephen Drury and his ensemble will premiere and record *for Lou* in Boston next fall. In the coming weeks I hope to complete the piece before immersing myself entirely in *The Place*.

JANUARY 17

I've been dreaming and thinking about *The Place* for the past five years. But its scale and complexity are daunting, as I face the intimidating question: How do I begin?

The answer came from my drywall man, Dave. Yesterday Dave came in to begin taping, applying corner bead and mud, sanding and painting the interior of our house. It's a huge project for one man, with countless details. But Dave just walked in, laid out his materials, opened his toolbox and started working.

When I asked him about this, Dave told me that inexperienced rockers sometimes walk around a new job and try to sort everything out in their minds before beginning work. This can be overwhelming, leading to paralysis. In his fifty years of doing masonry, Dave has learned that the best way to learn the details of a particular job is by doing the job. So he just begins.

JANUARY 20

The omnipresent sonic atmosphere of *The Place* will be harmonic fields associated with night and day, light and darkness. As I move deeper into this work, I find myself waking up each morning wondering: "How would this light, how would this weather *sound?*"

My hope is that this new work will deepen the listener's awareness of the world around us. Last night my wise wife, Cindy, observed: "In *The Place* we should learn more about what's going on outside than we knew when we entered."

FEBRUARY 2

As I put the finishing touches on the score, it occurs to me that *for Lou Harrison* completes a trilogy that I didn't know I was writing. This is my third full-length orchestral work composed in memory of a loved one. I composed *Clouds of Forgetting, Clouds of Unknowing* (1991–95) for my father. *In the White Silence* (1998) was composed for my mother. And now comes *for Lou Harrison,* in memory of the man who was a musical parent to me.

FEBRUARY 11

for Lou encompasses the most saturated textures in my music to date. But in form it's the simplest of the trilogy. *Clouds* is composed from four different musical textures. *White Silence* contains three. In *for Lou,* I've pared it down to two. Rising arpeggios sweep out of sustained harmonic clouds. And the quartet floats long solo lines over slow, procession-like music in the pianos and orchestra. These two textures alternate in nine continuous sections, each of which is grounded in a different five-, six- or seven-tone harmony. The formal structures recur from section to section throughout the piece. But the sound of the music is always changing.

FEBRUARY 18

Dirk Lummerzheim has become one of my principal science advisers for *The Place.* Tonight Dirk and I drive out to the Poker Flat research facility, where we spend the evening touring the facility, watching the aurora and talking about science and art.

I'm delighted not only by Dirk's sheer intellect but by the creativity of his thinking. The parallels between what he does and what I do are intriguing. Central to both our disciplines is the painstaking process of sorting through countless possibilities, trying to articulate the right questions. There's one essential difference, though. In Dirk's work there can be right answers. In my work there are none. Although I may solve compositional "problems" and arrive at solutions that feel right, final answers and ultimate meanings are always beyond my reach.

I'm deeply immersed in the science of *The Place* and enjoying it thoroughly. But my challenge is to make this more than a mere "science project." *The Place* is not a demonstration of natural forces. This is a work of art. Ultimately it will succeed or fail on its own terms, as art.

MARCH 7

Lou passed away in the spring of 2002. Not long afterward I dreamed I was rehearsing a new piece for chorus and gamelan. When I woke up, I wrote down the musical fragments and images I could remember from the dream. I was convinced this was the memoriam I would compose for Lou.

I played a while in the gamelan during my student days at Cal Arts, and I participated in performances of Lou's gamelan music when he visited Alaska. But that's as far as my experience goes. I've never composed for the medium, and in the months following my dream I came to feel it would be presumptuous for me to compose a gamelan work in memory of the master of American gamelan. So as the new instrumentation emerged, the gamelan dream faded . . . or did it?

This evening I played through one of the processional sections of *for Lou*. Suddenly it struck me that the interlocking layers of repeated melodic cells, the longer phrases punctuated by gonglike octaves in the low register of the piano, the stately pacing and solemn tone of the whole thing sounded a lot like Javanese gamelan. Trick on me! Despite my resolve not to invoke Lou's musical world directly, it struck me that I'd been unwittingly seduced by its charms. I laughed out loud. And in my mind I could hear Lou's hearty, joyful laugh.

MARCH 15

This is the season of change. Winter is moving toward spring. The weather fluctuates from day to day, sometimes wildly. Yesterday was warm and golden. Today is raw and gray.

The earth is moving more rapidly in its orbit. We're gaining seven minutes of daylight each day. At the other end of the year, as autumn moves toward winter, the rate of change is the same. But the feeling is different. Then the mood is one of rest or resignation, tinged with sadness. Now there's a palpable euphoria in the air. This is the time of new life and the ecstatic sense that comes with the return of "the light that fills the world."

MARCH 24

I'm thinking about the scale of *The Place,* in space and in time. In temporal terms this work is large. It's continuous, it never repeats and it never ends. But the physical space of the work is more intimate. It exists within a relatively small setting. So I'm wondering: How will the sense of time and space in *The Place* extend beyond the physical dimensions of the room? How can I evoke a sense of vastness in a small space? Can I find what Gaston Bachelard calls "intimate immensity"?

APRIL 10

Cindy and I are in New York City. Shortly after we arrived, we attended a concert of my music at an art gallery in SoHo. In the two days since then I've been visiting sound galleries here in Manhattan. I've enjoyed a lot of what I've heard, but without exception all these sound works are loud, continuous and presented in dark rooms. Amid the constant visual and aural overload of New York, retreating into a darkened cave may be the only way to focus on something other than the sensory-saturated environment of the city. (Everywhere you look here people are walking around absorbed in their cell phones, earphones and personal electronic devices.) Yet the deeper we retreat into our electronic caves, the more uninhabitable our world becomes. I can't help but wonder: What might happen if we turned our attention *outward,* as well as inward?

APRIL 12

Today at Paul Zinman's studio we had the final mastering session for *Strange and Sacred Noise.* Almost six years after we recorded it in Cincinnati, the piece sounds fresh and stronger than I remembered. It was a special joy for me that Al Otte was there to share this occasion. Our collaboration on *Noise* was one of the richest experiences of my creative life. It was also the point of departure for my continuing work with noise as a primary source for music. Although the media and the sounds are quite different, the route from *Strange and Sacred Noise* to *The Place Where You Go to Listen* is clear and direct.

APRIL 13

My dear friend Fred Peters is a diehard opera buff. Yesterday Fred took me to see *Die Walküre* at the Metropolitan Opera. I'm less interested in individual opera singers than I am in Major League pitchers. But this was, in Fred's words, "the cast of a lifetime." The singing was so strong it was transparent, and transcendent. In the third act, when Wotan put his favorite child, Brunhilde, to sleep amid the "magic fire music," I was almost moved to tears.

Wagner was a genius. But it's no disparagement of his achievement to say that he also caught the perfect wave. The social, economic and cultural currents of nineteenth-century Europe converged and crested just at the moment that Wagner emerged, giving him access to resources far beyond the wildest dreams of most composers today. Which makes me wonder: What waves may be rising in our time?

APRIL 16

I'm in Massachusetts for a brief residency at Williams College. Yesterday my host, the composer Bonnie Miksch, took me to the Massachusetts Museum of Contemporary Art, where I wanted to hear the permanent sound installations and to visit the large temporary installation by Ann Hamilton with sound by Meredith Monk.

Today I've had several sessions with composition students, coached a rehearsal and given a reading from *Winter Music*. The culmination of this busy day was a concert this evening, including a performance of *Red Arc/Blue Veil*.

APRIL 17

On my way back to Manhattan, I hop off the train to meet Frank Oteri at the Dia Beacon. This is my first visit to the new museum, which is dedicated to some of my favorite artists: Richard Serra, Donald Judd, Dan Flavin, Agnes Martin, Robert Ryman and Gerhard Richter.

As I often do in response to visual art that moves me, I imagine how the art might sound. In the Ryman rooms I hear soft white noise, floating in lush, enveloping veils. Inside the Serra sculptures I hear dark rumbling clusters, like the earth turning in sleep.

The real surprise of the museum is the string sculptures by Fred Sandback. With only a few wispy strands of yarn, this artist I've never before encountered creates magical force fields. The works are hardly there. You can see right through them. You can walk right through them. And yet you tend to walk around them, as if repelled by a powerful magnet of the same polarity as your body. When you finally summon the strength to pass through, there's a vivid sense that you've somehow crossed a threshold into a fractal dimension between dimensions.

Sandback does this with next to nothing. And I love the matter-of-fact way he states his aspiration to "assert a certain place or volume in its full materiality without occupying and obscuring it."

If only I could find a similar energy in musical space and place.

APRIL 18

Tomorrow we will fly home. Today Cindy and I take in a baseball game at Shea Stadium. Baseball is such an elegant game, so rich in precision and detail. Whenever I go to the ballpark, I keep a scorecard. Keeping score keeps me in the game. It's a kind of meditation aid that helps me stay focused on the present moment and the unfolding intricacies of the game. At the ballpark as in my life's work, scorekeeping seems to be what I do.

I love pitching. I love defense. I love base running. I love hitting. I love the strategy, the continuing calculations of the probabilities and possibilities of each situation. But my favorite thing about baseball is the simplest thing of all: the arc of the ball. That short interval between the pitcher's mound and home plate is so powerfully charged. And the arc of a high fly ball is a mesmerizing sculpture of time and motion.

APRIL 20

I've just returned home to learn that *The Place* has been moved. It seems there isn't enough funding to construct the space we've been designing. So the project is being moved to a space originally designated for a tiny office and an alcove off the main art gallery. This is one of the busiest locations in the museum, directly above the main entryway, near the convergence of two major stairways. I'm frustrated and distressed by this change. Rather than compromise the work, I'm seriously considering withdrawing from the project.

APRIL 21

As I ponder what to do, it occurs to me that if someone gave me a room somewhere, a workable budget, and asked me to create something special, I'd be delighted. This will be my attitude. I will embrace this as an opportunity. I will find a way to make the new space work.

So I'm back to square one. *The Place* may now be different from the work I've been imaging. But if I stay open and pay careful attention, the work will be strong.

APRIL 23

Maybe *The Place* shouldn't be dependent on the room. Yet how can it not be? My conception of the work must be broad and strong enough to make any inadequacies of the setting disappear. Yet the specificity of the work must grow from the specificity of the setting.

This piece needs to be created within the space. It needs to embrace the challenges and the possibilities of the space. Construction of the building is already well under way. So my job now is to survey the circumstances and exert whatever influence I can on the completion of the room. I'm told I'll be able to move into the space by next fall. From then until the opening (scheduled for the following year) this will become my studio.

Robert Irwin speaks of working with "anything that references against the specific conditions of the site." As he puts it: "Whether it works is my only criterion."

The new location is noisier. It's situated directly above the busiest corridor in the museum. This suggests that the sound world of *The Place* should be fuller and more sustained than I've been imagining. Instead of the elliptical space I'd envisioned, the new space is narrow and much smaller. There won't be much room for either the sounds or the listeners to move around. I'll probably cover the large bank of windows on the south side of the room with some translucent material. Even so, the windows will still define a strong sense of front and back to the room.

In response to all this, the work needs to be simpler, more concentrated. My conception may remain orchestral, but the listening experience will be more intimate, more refined, more like chamber music.

MAY 16

Last week we passed that day when the air sounds as it does on no other day of the year. Here at sixty-five degrees north latitude, the leaves come out quickly. Almost overnight we move from the sound of wind on bare branches to the sound of rattling leaves. But on that one day the air has a special softness, the sound of the forest breathing through half-opened leaves. The leaves are now fully unfurled. I relish these days of constant light. And I cherish the feeling of settling into a steady rhythm of work.

It's a great privilege to live as an artist. It's also a responsibility and a discipline. Without the routines of conventional employment, an independent artist must create his or her own structures for life and work. The work is difficult enough. And it's a constant challenge to resist distractions and busyness.

As part of her discipline, Agnes Martin reminds herself daily: "I have everything I need to do my work." This is always true. For me now, this is true as it's never been before. I'm still a bit overwhelmed by the size, complexity and unfamiliarity of *The Place.* But I have everything I need to do this work, including time. There is no rush. Right now I'm laying the conceptual foundation of the work. This is a crucial stage of exploration and discovery. I need to enjoy the process, trust the work and let it emerge in its own time.

MAY 17

This afternoon, I walk up the mountain above the studio. This is my work today: walking. It slows things down and quiets my mind.

To the south the Tanana Valley sprawls out to the peaks of the Alaska Range. To the north I look toward the view you can't quite see . . . to the Yukon Flats, to the great river itself, to the Brooks Range and the Arctic coast beyond. This is the country of my dreams, the landscape of my soul.

Back in the studio, I realize that I've lost my notepad somewhere up the mountain. So I sit down and write everything out from memory.

MAY 18

Today I walk back up the mountain. Along the way I find my notepad, not too worse for the wear from last night's rain. I pick it up, and with it the thread of yesterday's thoughts. I continue on to the top.

In some Buddhist traditions the monks practice walking as a form of meditation. This makes sense to me. Years ago I asked a psychologist friend whether he thought I should undergo analysis. He smiled, looked me in the eye and said: "John, the best thing you can do is to walk."

MAY 19

At the moment I'm sketching, studying and thinking. But soon I'll need to begin working with sound. It's time for me to assemble my tools and materials. I feel a bit like the Yup'ik subsistence hunter: "always getting ready."

I've ordered a large piece of scrim material and various types of diffraction and diffusion coatings for glass. As I wait for these to arrive, I'm pondering white noise as the aural equivalent of white light. The full spectrum of visible colors is contained within light. Is there a similar spectrum of audible colors hidden within noise?

MAY 20

If one visitor to *The Place* experiences something meaningful or glimpses something new, I'll be pleased. If another visitor falls asleep, that will make me happy as well.

MAY 22

I should keep walking. This afternoon I walk back to the top of the mountain. On the way up I think through the details of my current work, trying to wrap my mind around the most fundamental questions about *The Place*. As I stand at the top, looking over the country stretching away all around me, things become clear.

The Place will be composed of three elements: light, noise and tone.

The primary sonic element will be colored noise derived from the changing angles of the sun. I imagine the sun as a giant noise generator, or a cosmic tam-tam. As the sun rises and falls, the fundamental pitch of the tam-tam and its noise will follow. Below the horizon the noise will fall, reaching the lowest audible frequencies at solar midnight on the winter solstice. Above the horizon the fundamental will rise, reaching its highest frequency at solar noon on the summer solstice. The noise will move around the space according to the movements of the sun above, below and around the horizon.

Just as the visible light in *The Place* may be filtered through a scrim, the noise will be filtered through two "harmonic scrims." These may be tuned to the harmonic series, or perhaps to the natural resonances of the room.

Other sounds will be derived from seismic data and magnetometer data. These sounds will be very low frequency and very high, respectively. They will encompass narrower frequency bands than the sun noise and tuned to their own harmonic colors.

The magnetometer sounds will be localized high in the room, and the seismic sounds will rumble the floor. Both may be distributed around the horizon according to the direction and distance of the stations from which the data are derived.

MAY 24

There are few musical models for my current work, few influences that seem directly pertinent. Most of my influences have long been assimilated into my music. I'm on my own now. Most of the time I feel as though I'm working (as Richard Serra says) from out of my previous work.

From time to time I've turned to Native cultures for inspiration. In recent years I've also turned to visual art, searching for musical equivalents of color and texture, space and presence. But it seems that, sooner or later, everything leads me back to nature as my primary source. Nature is geography and geology. Nature is biology and ecology. Nature is physics. More specifically for a composer, nature is acoustics—the physics of sound. Nature is inexhaustible.

MAY 26

Like silence, noise is both a rich metaphor and a powerful sounding element. Noise will be the *prima materia* of *The Place*.

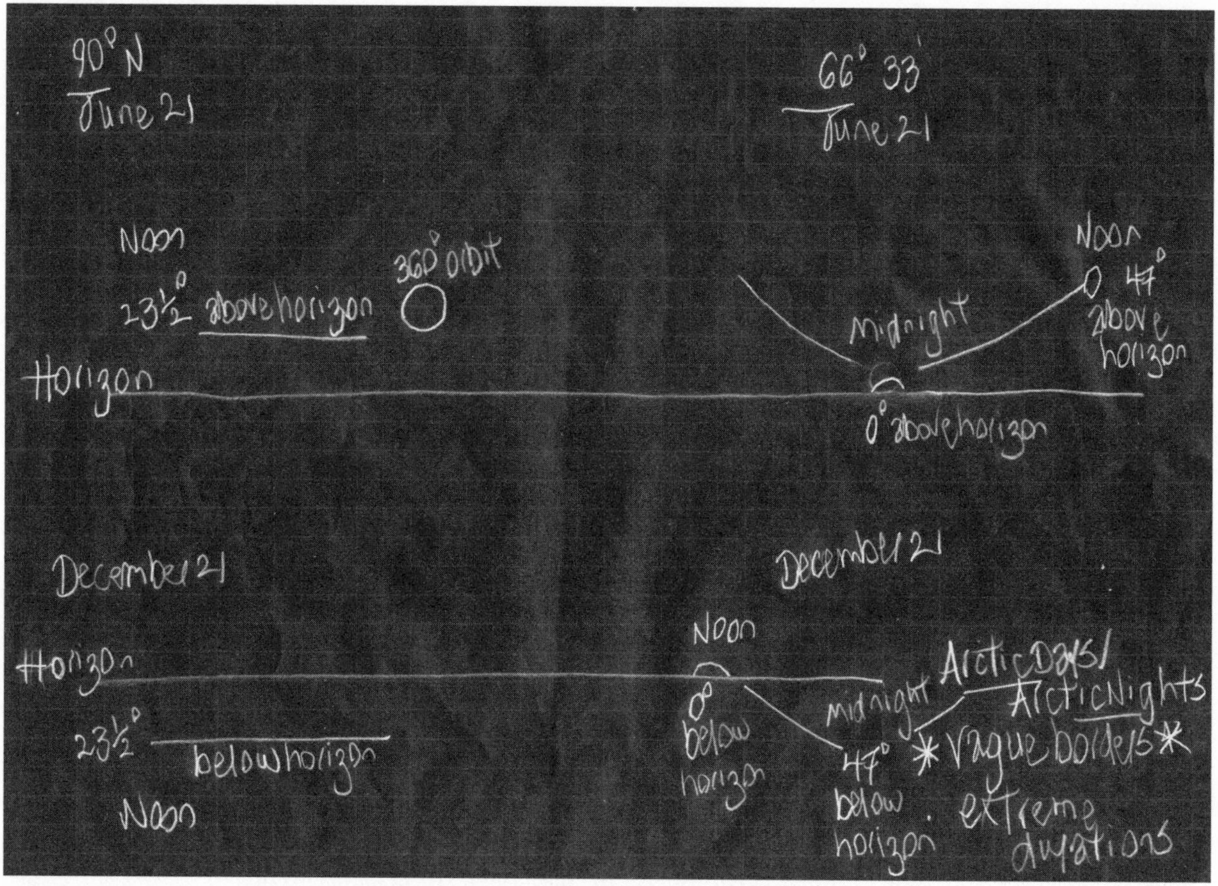

Figure 1.
The sun in
the Arctic sky.
(Composer's
sketchbook.)

Noise contains worlds of possibilities. In this new work I feel a bit like the Kaluli song-maker who camps by a waterfall until he hears his new song within the roar of the water. My job is to listen for these inner voices, the essential tones within the noise of the world.

The deeper I move into the world of *The Place,* the more I'm coming to understand and to trust the process it demands. My anxiety is dissipating. Instead of worrying whether I'll have anything to paint, I'm now embracing the emptiness of the canvas. I want to leave the material untouched for as long as I can.

MAY 31

Cindy and I have just returned from several days in the mountains. We met our buddy Gordon Wright for our annual spring camp at our favorite spot in the Alaska Range. To be on the tundra with the nesting birds and the first flowers of the season felt like coming home after a long time away.

JUNE 14

This morning I happened to flip on the radio just as *Prelude to the Afternoon of a Faun* was beginning. I sat down and listened. There were moments when I just wanted the music to stop, to freeze in time. And I wondered: Could I actually *do* something like that? The performance was a good one. The orchestra sounded rich and supple. It also sounded archaic. And it struck me with new force that this is not and cannot be my primary medium. On occasion I may have access to the orchestra. When I do, I will embrace it. But electronic media offer me rich and supple possibilities for creating new worlds of sound. This is my new orchestra.

I've always loved Debussy. But the distance between his world and mine seems greater than ever. I feel as though I'm now listening from beyond the outer edge of "classical" music. It's a good place to be. It feels like home.

JUNE 17

Last night I attended an outdoor performance of *The Winter's Tale.* This was a good amateur production. Several of the actors were quite convincing. But I found myself less captivated by the drama than by the dynamism of the setting and the sheer music of the language.

The wind blew. The sky was constantly changing. When the ambient noise made the words difficult to hear, somehow their meaning became clearer. Fragments of human speech mingled with the song of a white-crowned sparrow and blew away across the open fields.

In an outdoor setting, we're forced to integrate a broader range of sensations. Our sphere of attention expands. We perceive things differently.

Can we bring the sense of openness and wonder we feel in nature into the interior spaces in which we usually experience art? And can we carry the sense of expectation and attention that we bring to the theater or the concert hall out into the natural world?

JUNE 21

On the summer solstice I've begun playing with noise, making veils of sound from slow descending sweeps of filtered noise, superimposed layer upon layer. I'm not sure exactly what I'm doing or why. But it feels as though I'm in touch with something elemental.

JUNE 27

In the sub-Arctic summer, it's light all the time. So this evening about 1:00 A.M. Cindy, our friend Annie Caulfield and I drive north, toward the sun. After ten days watching smoke clouds billowing on the horizon, we decided to visit the Boundary Fire, which is currently burning about forty thousand acres of boreal forest off the Steese Highway.

As we crest the hills north of Fairbanks, the sky glows with the color of wild roses. From high above, long fingers of virga extend downward—the rain reaching toward, but never quite touching, the earth. In the west a half-moon hangs, impossibly large, yellow-gray.

We descend into the valley. The sky darkens. The ground is barren. The trees are black. Banners of smoke rise from the slopes above us. A fine mist of ash begins to fall. As we move deeper into the country, flashes of orange light appear, one by one, like a hundred campfires scattered all around.

We stop the car. Amid the flames and smoky light, the birds are still singing. The music of hermit thrushes surrounds us like the echoes of small bells, mingling with the murmurs of a stream rising from the darkness below. And another sound—first breathing, then sighing, then clattering like stones rolling down a talus slope—as a geyser of flames erupts from the crown of an ancient spruce.

We walk to the top of a steep draw and find ourselves standing on the edge of an inferno. A searing wind roars up the slope, showering sparks into the sky. It is like standing on the bank of a dangerous river. With one more step we could be swept away, consumed by the rising torrent of fire. We're frightened. But we don't move. The heat is intense yet somehow comforting. The sound is ominous and mesmerizing. The fire pours downslope, devouring the forest, moving inexorably toward the river.

We drive on, descending toward the Yukon River. Gradually as the slopes fall away we can see the edge of the smoke, hanging like an enormous black knife low on the horizon. Beyond, to the north, pale blue light stretches away into endless morning. A gentle rain begins to fall.

I know that fire is a central element of this ecosystem. I often speak of wildfire as an example of a force that appears to be catastrophic but that actually furthers health and growth. Yet this year's fire season is the worst I can remember. It's hard not to imagine this is a result of climate change.

JULY 5

The fires have grown. We've still had no rain, and the prevailing northeast winds these past several days have fanned the flames and brought the smoke in. We're wrapped in a noxious pall. The sun has disappeared. Visibility is near zero. The stench is overwhelming.

JULY 7

Today was a good day in the studio. After several days trying to find my way back into *The Place*, I had a productive session working out the arcs of the noise fields as they rise and fall with the sun over the course of the year.

At one moment I thought: "This work is so difficult because I have to make everything up." Instantly I realized, "No. That would be easy! This work is difficult because it has to be *real*."

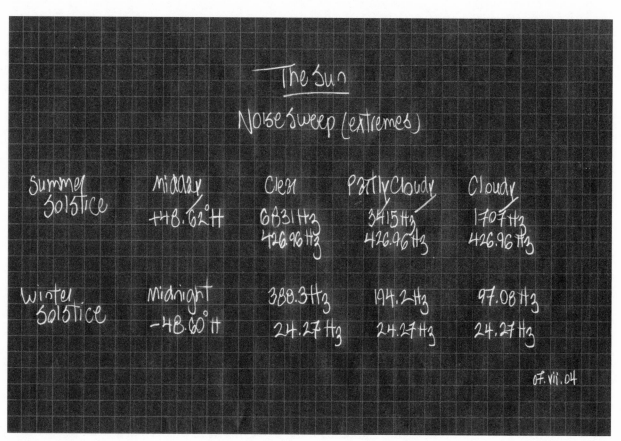

Figure 2.
The noise sweep of
the sun in The Place.
(Composer's sketchbook.)

My challenge is to listen to the world around me, to hear things that haven't been heard before and to imagine how those things might sound. I want *The Place* to feel like a world of its own. Yet this self-contained musical world is all about tuning our ears to the larger world in which it exists. This work is so difficult because I can't just make it up. It must resonate not only with itself, but also with the natural forces that animate it. *The Place* must ring true.

I often work with compositional "systems" of my own invention, formal processes and techniques I use to help me find the music, to give it intellectual coherence and sensual presence. In *The Place* the controlling systems are not simply musical. They are geophysical. The moment-to-moment continuity of this music is controlled by physical phenomena in the real world. In a sense, the forces of nature will make this music.

JULY 11

What am I doing, spending these long summer days playing with noise? It's difficult enough to justify composing music these days. Yet this is what I've always felt called to do. Composing is my life's work. It's all I really know. But these days I'm not even sure that I *am* composing. So here I am with no clear idea of what I'm doing or why I'm doing it. Yet somehow it feels like I'm doing the right thing.

The noise itself isn't even real. This is not the noise of ocean waves or a waterfall. It's artificial noise, produced on a computer by random number generators. Still, I feel as though I'm working with something as powerful and elemental as fire. Deep within the roar of these waves of noise are hidden all the songs and all the voices of the world.

JULY 15

Working only with noise, I feel alternately like a child playing with matches and a primitive man who has just discovered the magical powers of fire.

Fire is an all-too-real presence these days, as wildfires are still raging all over central Alaska. The trees are desiccated. The combination of leaf-miner caterpillars, lack of rainfall and unremitting smoke is more than the birches and aspens can withstand. They've begun to turn yellow. And these past few days the wind has risen. The ground around the studio is now covered with leaves. It's only mid-July. But it looks like late September.

JULY 21

A new sound is coloring the air in my studio. A new kind of harmony seems to be emerging from these fields of colored noise. With each passing day I'm learning more about the nature of my new materials. I've been listening to different bandwidths of filtered noise, each of which produces a different weight and texture of tone. I'm exploring a continuum from single, breathy tones to broad bands of colored noise.

After trying out scrims with graduated bandwidths (wider and narrower over the audible spectrum), I've settled on scrims of uniform porosity from bottom to top. This produces a fuller sound, with sounds evenly distributed throughout the tonal space. As I explore these different varieties of noise, I'm relishing the sense of physicality in the sounds and this sculptural way of working with them.

JULY 23

The entire Yukon River basin—from the Canadian interior to the Bering Sea—is filled with smoke that airline pilots have observed now extends to an altitude of twenty or twenty-five thousand feet. How many hundreds of thousands of cubic miles are submerged in this vast ocean of smoke? And here we are in the forest and on the tundra, trying to breathe at the bottom of the sea.

AUGUST 2

Day to day I live as though faith and life are two different things. Faith, so my thoughts and words and actions seem to say, is an ideal. Life, the same thoughts and words and actions seem to say, is something else. But faith that is only an ideal isn't really faith. And though my faith doesn't have a recognizable name, it's a very real part of my life. For me, the practice of music is a practice of faith. Composing is an act of devotion. Can I find a way to bring the mindfulness I bring to my work into every moment and every dimension of my life?

AUGUST 7

This feels like a siege. Day after day we've had air quality alerts, warning people to limit their physical exertion and stay indoors as much as possible. The Fairbanks hospital and the Red Cross have set up a smoke respite station at the hospital to give people sanctuary from the oppressive conditions.

If this were happening in Los Angeles or New York, it would be considered a major emergency. It would be headline news. But there's almost no coverage of this northern inferno in the national news media.

AUGUST 10

For me art is a form of prayer. As the wildfires continue to rage, I remember the Zen master who taught: "Always pray as if your hair were on fire."

AUGUST 18

We've just returned from a week camping in the upper Sheenjek River, just south of the Arctic Divide. What a joy it was to be in the Brooks Range again. Those mountains are where my spirit feels most at home. The weather was unusually hot. But what a relief it was to escape from the smoke!

The company was the most agreeable: Cindy and I, Gordon and our friend Edith Rohde, and Fred and Alexandra, all the way from New York. We had no plans. Each day we followed wherever our inclinations led us, hiking up the valley, down the valley, looking for animals, soaking in the luxurious space and light. One particularly warm afternoon Fred and I went skinny-dipping (briefly!) in the bracing waters of the Sheenjek.

The highlight of the trip for me was my climb up an unnamed peak on the east side of the rugged valley. After all these years it's still one of my greatest joys to scramble up the rocky slopes, survey the sweep of the country below and listen to wind singing over the summit.

A Composer's Journal Part II—Studio Notes

SEPTEMBER 1, 2004

Usually my work is solitary. But suddenly I find myself sharing the studio with someone else. Jim Altieri has just arrived from Thailand to work with me for a couple of months on *The Place*. Jim was one of my favorite students when I taught at Oberlin, and it's a joy to welcome him here. Jim is a brilliant programmer. He has a background in science and a keen musical sensibility. A composer himself, Jim understands the nature of the work ahead of us. He also has a delightful sense of humor and is a pleasure to be around on a daily basis. I spent most of my younger years in relative isolation. Now, in middle age, I find that the occasional company of younger musicians helps keep me in touch with my own idealism and faith in music.

SEPTEMBER 3

I'm accustomed to composing at some distance removed from the flow of time in the sounding music. But in recent months I've stepped even farther outside time. The sounds I imagine for *The Place* aren't available to me on existing instruments. So I've been searching for them through a slow and laborious process involving several different computer programs.

Now that Jim is here we've leaped directly into the world of real-time processing. This is exciting new territory for me, a way of working that I've dreamed of for quite some time. It's very satisfying to be able to sculpt sound directly and hear the change immediately. And it's thrilling to hear data derived from natural phenomena—the changing angles of the sun, perturbations of the magnetosphere and seismic activity—transformed into sounds that trace the contours and rhythms of these phenomena.

Jim and I find ourselves absorbed in details of the sound the way we might be absorbed in the minute features of a tundra landscape. A difference of 10 cents (one-tenth of a tempered half-step) is audible and significant.

SEPTEMBER 7

I've always loved maps. This probably comes from my love for the land itself. A map is a symbolic text that can give us new viewpoints and deeper insights into the lay of the land, the topography of its mysteries. The lines of a map constitute a drawing, a score that evokes the music of the land.

I'm a fairly good reader of maps. On camping trips I always carry them with me. A good map can deepen my understanding of exactly where I am. Occasionally I've used a map to help me find my way in unfamiliar terrain. Now in my work I'm using maps to help me find my way into this strange new sonic terrain.

Jim and I are mapping data. We're taking streams of numbers that represent geophysical phenomena and using them to create maps of another kind. From these musical maps we're generating sounds—derived from the original phenomena, but very different in sensation. Sunlight and the colors of the sky become singing voices. Geomagnetic disturbances associated with the aurora become shimmering bell tones. The movements of the earth become rumbling drums.

The terrain becomes a map that yields a new map that reveals a new terrain.

SEPTEMBER 12

The Place has opened up a new world in my imagination, a world I may inhabit for years to come.

I've been thinking about a series of performance works drawn directly from *The Place*, to be performed within the space. These would be instrumental and choral transcriptions of specific days and times, at time scales ranging from 1/60 (one hour accelerated into one minute) to 1/1 (real time).

I've also been dreaming about a future installation work translating weather and climate data from all over the earth into sound and light.

SEPTEMBER 25

Neither classical music nor commercial music feels like home to me. My music doesn't belong to either of those worlds. And I wouldn't want to choose one or the other any more than I'd want to choose between the socioeconomic systems they embody. Neither Old World despotism nor global capitalist monoculture embodies my ideals. I'd rather be a citizen of an emerging world that doesn't yet exist, except in the minds of those who imagine it and help create it.

OCTOBER 3

Long before Europeans came to northwest Alaska, Maniilaq, the great prophet of the Kobuk River, predicted there would come a time when a season would occur twice in succession. Ever since then most people have believed this prophecy would come to pass as two consecutive winters. But Maniilaq did not say. And today it seems more likely to come as two summers.

For those of us who live here on the northern line, climate change has become an undeniable part of our daily lives. During the thirty years since I first came here, Alaska has been warming. Winters have become milder and shorter. The summer wildfire season has increased in length and intensity. This year it lasted all summer.

Following the wettest, cloudiest May on record, there was virtually no precipitation for the rest of the summer. The average temperature for May through September was the warmest on record, exceeding the previous record by a least 1°F. This was the worst fire season on record. A total of 6.6 million acres of Alaska (an area larger than the state of Vermont) went up in flames.

How will long-term changes in climate affect the boreal forest? Will the spruce, birch and aspen give way to vast grasslands? Some animal species will adapt and survive. Others will vanish. Might we humans be among them?

Maniilaq also predicted there would come a day that would suddenly appear to split in half. When the people asked him what would happen after that day, he refused to speak. He would not speak of any future following this day.

OCTOBER 7

Wittgenstein observed: "Looking does not teach us anything about the concept of color." Perhaps. But the concept of color does not teach us anything about the experience of looking, about the *sensation* of color.

We are not brains hermetically sealed in jars. We are corporeal beings living in an animated sensual world. Our thoughts and our sensations are one and the same.

OCTOBER 16

As Jim and I move deeper into work with the seismic and geomagnetic data, we're encountering tricky little math problems because we're manipulating the data in real time. Most of our science advisers analyze their data after it's been gathered, rather than working with it as it appears. So as we talk with them about how to approach this, even they are scratching their heads a bit. As Jim quips: "There isn't much real-time science."

Science is grounded in empirical analysis of observed facts. Art, too, is grounded in observation. But the observations of art are more phenomenological, less about analysis and more about perception. Rather than describing phenomena, art provokes experience.

OCTOBER 24

Following the summer from hell, winter has finally arrived. The snow never looked so good. Some of the fires are so intense that they may survive the winter. But at last the air is beginning to clear.

Jim has left for the winter, and the stream of visitors we've had these past few months has subsided. I'm looking forward to the dark days of winter to regain my balance, to reconnect with the core of my work. I feel the work opening up, expanding from the local to the global. For years I've explored the metaphor of sonic geography. Now I sense this expanding toward sonic ecology.

NOVEMBER 2

On election eve I find myself praying for this country. May the people speak with wisdom and compassion. May our body politic include future generations, all other species and the earth itself.

This afternoon I heard a news report about a study on climate change that's just been released. Although it won't capture the headlines, this story may ultimately prove to be more significant than the results of today's election. The study concludes that large-scale climate change is under way and that the rate of change will continue to accelerate unless decisive steps are taken to slow it down. Warming is occurring twice as rapidly in the Arctic as in other regions of the earth. Over the past thirty years the thickness of the ice cap has diminished by 50 percent. As soon as 2050 there may be no multi-year ice in the Arctic at all. By the end of this century polar bears may be extinct.

For me, this is personal. This is my home. I love the North with a passion that is physical. I feel my own body and soul as extensions of this place. I came here as a young man pursuing the possibilities I imagined here. The Arctic was our geography of hope, our last great, untouched place. But the reality that confronts us today is starkly different. The Arctic has become the most threatened part of this threatened planet.

As goes the Arctic so goes the earth. And as goes the earth so goes the human species. If the Arctic remains our geography of hope, it no longer holds the promise of the pristine. The promise of the Arctic today is as the first place in which we discover for ourselves a new consciousness, a new way of living on this earth.

NOVEMBER 3

The outcome of the election seemed inevitable. Polls show that a wide majority of Americans feel that in the last election the president was not duly elected. Yet apparently the voters have ratified that election with this one. We're being told that voters' sense of "moral values" prevailed over concern about war, the economy and the environment.

This evening I heard about a candidate in Oklahoma who apparently won because he took strong "moral" stands, opposing homosexual marriage and advocating death for physicians who perform abortions. What kind of twisted morality is *that*?

Whether Islamic or Christian, Arab or American, fundamentalism is an ugly, frightening thing. Yesterday's election compels me to renew my commitment to my work and to do whatever I can to resist ignorance and hatred in this world.

NOVEMBER 4

When I began working with the magnetometer data, I wanted to orchestrate the data to sound like the aurora looks. I imagined audible forms that would move in synchronization with aurora forms in the sky.

Gradually it's dawned on me that the magnetometer data and the aurora are two different phenomena. It's true they're closely related. When the magnetometer readings are active, so is the aurora. And vice versa. But it's virtually impossible to find specific correlations between geomagnetic activity and the visible forms of the aurora. As Dirk observes, the magnetometers don't measure the aurora. They measure "space weather."

This past week a major geomagnetic storm occurred. People in the temperate zone have been treated to spectacular displays of aurora. Here in Fairbanks, however, we've had snow and dense cloud cover. So we haven't seen any of this activity. Inside *The Place* we would have been able to *hear* this storm.

The Place doesn't illustrate the visible. It doesn't amplify the audible. It resonates with the inaudible and the invisible.

NOVEMBER 11

Last night in the studio my composition student Julian Darpino asked me: "How are you working on this piece right now? What exactly are you doing?"

Good question. Last summer I found myself asking the same question as, day after day, I played with colored noise shaped by the rhythms of day and night. Even though I wasn't always sure exactly what I was doing or why I was doing it, my summer work laid a solid sonic foundation for *The Place*.

This winter I'm immersing myself in the other two primary elements of this music— the Aurora Bells and the Earth Drums. I'm listening. As I listen, I'm refining the details of my mapping schemes for translating the forces of geomagnetic weather and seismic activity into sound. Just as I would do with a work for human performers and acoustic instruments, I'm giving it the test of time. I'm considering and reconsidering the instruments, the sounding materials and the forms of the music. Just as I would do with a new piece in a more familiar medium, I'm working with harmony, melody, duration, orchestration, texture and other musical elements. As I sit listening to the interplay of the geomagnetic bells above the dark drone of the night, I can imagine this music transcribed for acoustical instruments.

So it turns out that I'm doing what I've always done. I'm composing.

DECEMBER 1

Some composers seem to think that music is the whole world. But I'm more inclined to believe that the whole world is music.

DECEMBER 13

I often speak and sometimes write about "truth." But what do I mean by that word?

I avoid using the noun with an article. To speak of "the" truth would imply belief in an absolute quality that I don't ascribe to truth. When I use the term, it usually implies "truth as I understand it." But to speak of "a" truth or multiple "truths" would imply an easy relativism that I also don't ascribe to truth. This would place a distance between truth and my own commitment to it.

Truth arises from our deepest desires and most honest efforts to understand the world in which we live. Inevitably those desires and efforts involve us in language and thought. But our search for truth precedes and transcends language and thought. Truth does not exist in a vacuum. It arises only as we participate directly in the mystery of life. Only when we are fully engaged with existence, willing to risk everything we know and everything we are, can we experience in occasional fleeting moments the full presence of truth.

Maybe it's not truth itself that we hear, but only the reverberations, the resonances, the *ring* of truth. Perhaps that's as close to truth as we can get.

DECEMBER 21

Very late this morning (at least by the time of the clock) I woke to deep golden light glowing on the wall opposite our bed. Rising and walking upstairs, I watched the burning disk of the sun float slowly up out of the dark gray clouds hanging over the mountains.

Today is the winter solstice. To me this is the holiest day of the year. The midwinter season can feel like falling into a deep tunnel. I used to approach it with a vague sense of dread. But now this is my favorite time of year. The deepest darkness is the birth of light.

The darkness seems to encourage more reflective, introspective thoughts. I call it "winter mind," and I savor this as the season to burrow in and spend long quiet hours in my studio. This is a time to ponder stillness and light, birth and renewal.

It used to be when people in other places would ask me about winter in Alaska I'd tell them: "The cold isn't so difficult. It's the darkness." But over the years I've come to love the midwinter darkness as much as the midsummer light. After all, darkness and light are two sides of the same experience.

JANUARY 6, 2005

We're in the midst of another January thaw. This one is accompanied by howling winds, and wet, heavy snow.

I heard a meteorologist on the radio today who said that in eighteen years of working as a forecaster in Alaska, he's never seen anything like the current conditions. An extreme high-pressure system is sitting over the Bering Sea. According to the weatherman this would be an unusually strong front even in summer. In winter, it's almost unbelievable.

The news is filled with reports of the devastation from the tsunami in the Indian Ocean. As the death toll continues to mount, I can't help but wonder whether this is some sort of horrific prelude to much greater disasters to come, disasters of our own making. It's heartening to see people all over the world uniting to help the victims of this present disaster. Now if we can only maintain this sense of unity and work together to do whatever we can to save all of us living on this beleaguered earth.

JANUARY 10

In Inuit tradition the force that animates all things is *sila,* the breath of the world. *Sila* is wind and weather, the forces of nature. But it's something more. *Sila* is intelligence. It's awareness: our own awareness of the world, and the world's awareness of us. If we listen carefully to the breath of the world, perhaps our music can become filled with this awareness.

I'm making more sketches for the projected weather piece. It will be called *Sila: The Breath of the World.* Like *The Place Where You Go to Listen,* I imagine *Sila* as an invitation to heighten our awareness and, also, to contemplate our influence on the world. This work will celebrate the complex forces of weather and climate, transforming data from stations around the earth into an enveloping continuum of tone and color.

Each station will have its own voice tuned to the breath of its geographic location. Heard together around the space of the work, these voices will create a virtual choir that sings with the breath of the world. Within *Sila* each station will be illuminated in its own wash of color. Seen together these colors will evoke a composite atmosphere of the whole earth. As the temperature at each station rises and falls, the voice of that place will rise and fall in pitch, and the color of light will become "hotter" or "cooler." Changes in wind and sky conditions will further modulate the sound and light.

Like *The Place, Sila* will be open-ended and unpredictable, unfolding in real time. The composer will design and tune the instruments, and shape the overall form of the work, but the world will provide the content.

JANUARY 16

It's 3:00 A.M. For the past hour and a half I've been watching the aurora. Tonight's forecast predicted maximum activity. I haven't been disappointed. As I write, every quadrant of the sky is blazing with red and green light moving in a breathtaking array of forms.

A few moments ago I saw something I've never seen in thirty years of aurora watching. In the western sky, a wall of red aurora shot upward, again and again, like a rising waterfall of fire. Usually we see the aurora in vertical forms that appear to move laterally. But these were horizontal forms moving vertically, rising continuously, sweeping up and up again toward the zenith. The speed and intensity of this display were almost frightening, like a vision of the apocalypse.

Now I'm afraid to go to bed, for fear I'll miss something. This is one of the maximum global electromagnetic storms that occurs every year or two. During the last one, people saw aurora in Utah and North Carolina. Here in Alaska, we didn't, because our cloud cover was so dense. Tonight is a different story. It's crystal clear and cold, about thirty below. The half-moon set about 1:30 A.M. Ever since then the aurora has been dancing.

At one point the light was so extensive that you couldn't really see it. There was too little contrast. The entire sky was saturated with phosphorescent green.

Earlier tonight working in the studio, I heard the boreal owl call for the first time this season.

This world is full of magic.

JANUARY 17

The storm continues. The sky is filled with ghostly white light. The weather is clear. There are no clouds. But the aurora is so dense that only the brightest stars and planets shine through the photon mist.

Today I spoke with my friend Dirk, the aurora scientist. He was also watching the aurora last night. When he went to bed, right around moonset, he figured he had pretty much seen what there was to see. When I told him about the wall of fire, he asked me to describe it in more detail. There was more than a trace of envy in his voice. I don't blame him. That waterfall of fire was a vision I'll never forget.

Earlier this evening Cindy and I watched for fifteen or twenty minutes as curtains and bands of red and green aurora danced all around the sky. Now it's 2:30 A.M. The moon is setting. But once again I'm reluctant to go to bed, hoping I might see another display of that fire in the sky.

A while ago, walking back to the house from the studio, I stood outside for a few minutes. Although it's nearly forty below in town, it doesn't seem too cold out here. And there's an absolute stillness to this night that (with the warmer winters of recent years) I haven't experienced in a long time. No wind. No animal sounds. No sounds of motors in the distance; only the sound of my own pulse, throbbing in my ears.

JANUARY 18

This evening Dirk gives a public lecture about aurora sound. I attend.

For centuries people have reported hearing sound in conjunction with active displays of aurora. Despite the varied times, locations and cultures of the observers, the reports are remarkably consistent. So it's hard to dismiss them out of hand. Yet there's no conclusive evidence for the existence of such a phenomenon. And there isn't a hypothesis that seems to explain it.

Virtually all observers have noted a close time correlation between the sound and the visual movement of the aurora. But since the aurora is at least a hundred kilometers above the observer (and usually some lateral distance beyond), it would take several minutes for the sound to reach the observer standing on the earth. Given the slow speed of sound, there is simply no way that such a sound could be located inside the aurora.

Some people have suggested that aurora sound is a function of "crossed wires" in the observer's brain. Somehow visual impulses that should register in the occipital lobe get routed to the temporal lobe. No one has explained why aurora should cause this when other visible phenomena don't. Still, I like this hypothesis. I can't help but imagine that aurora sound is a form of synesthesia, a result of the innate promiscuity of our senses and their tendency to respond sympathetically to strong stimulation.

To many scientists, the most plausible explanation for aurora sound is that people are somehow hearing ELF radio waves related to the aurora. However, the mechanism by which they might "receive" these signals is not known. So aurora sound remains a beautiful mystery.

After his talk, Dirk tells me that the rising waves of red that I saw the other evening were probably a rare form called "flaming aurora." I feel very fortunate to have seen it.

On my drive home, the aurora is dancing all over the sky.

JANUARY 23

Cindy and I are on retreat. We've come to this quiet lodge in the Alaska Range to mark my birthday, to reflect on the past year and to plan for the future.

I'm profoundly grateful for the blessings of my life and work. I want to live ever closer to the earth, with deeper awareness of the sentience all around us. And I want to follow the work wherever it may lead me for as long as I can.

JANUARY 25

It's 1:30 A.M. The moon is full. I sit watching stars, in the sky and in the snow.

The deep cold has pulled the flakes into large feather crystals that reflect the moon-light, creating points of light more brilliant than all but the brightest stars in the sky. Like the stars above, these snow stars form constellations. I watch them carefully. After just a few minutes, everything seems to have changed. I pick a particularly prominent snow star and sit very still, focusing on it intently. Before long this star has faded and others have appeared. These snow constellations change much more rapidly than the constellations in the sky.

The night is perfectly clear. The moonlight is not filtered through the trees. It falls directly onto the snow in the open field outside the house. I marvel at the constancy of change. And I wonder how this slowly turning field of stars might sound.

JANUARY 26

I'm watching the snow stars again. Last night I watched them fade. Tonight I want to see them appear. Choosing a section of the field containing no points of light, I watch until the first trace of light begins to emerge. At first I'm not sure it's really there. But over the next two or three minutes it grows to become an intense particle of illumination, burning in the snow. Another two or three minutes later, it's gone, dissolved back into the diffuse blue-gray whiteness of the field.

The entire life cycle of one of these stars is perhaps five or six minutes. With just a little patience I can witness the birth and life and death of entire galaxies. I look up to the sky and wonder about the scale of time concentrated in the points of light that touch my eyes from those faraway suns. The light from them comes from the time before time.

JANUARY 29

Last night, in my dreams, I held the wind in my hands. Or was it fire? Or was it some other primal essence? I stood on the summit of a naked peak. Between my outstretched arms I held a large ovoid shape that twisted and turned in my hands. As it changed form it also changed color—from red to yellow to orange, to brown to green to blue.

At first I was frightened. The thing seemed to move with a force of its own. It was powerful and clearly dangerous. The best I could do was simply to hold on and let it move as it would. But gradually I began to sense that I could gently influence its movement, shape and color. As I began to dance with it, I could feel its energy streaming into my body, giving me strength. Even so, I understood that this thing could easily destroy me. If I resisted too much, it could throw me from the mountaintop. I had to maintain a very delicate balance.

This dream is the way I understand my current work. In *The Place Where You Go to Listen,* I'm working with forces of nature more directly than ever before. The streams of seismic, electromagnetic and astronomical data that animate this work are mathematical descriptions of physical phenomena. They are not the phenomena themselves. But the complexity, dynamic range and time scale of these streams extend far beyond any images or metaphors I've worked with before. As with the strange force in my dream, my role in this new work is to move with the shifting forms, weights, textures and colors of forces that are ultimately beyond my control.

JANUARY 30

We can think about art. We can write and talk and argue about art. We can use art as a vehicle to convey our ideas and our beliefs. But at a certain point, art has to stand on its own. To support real meaning, art must first and foremost be itself.

FEBRUARY 9

Except in a vacuum, silence does not exist. Outside laboratory conditions, perfect sine tones and absolute white noise can't be found. The harmonic series is another elegant conceptual construction that we believe *should* exist. We hear portions of the harmonic series all the time, in our musical instruments and occasionally in the larger world around us. But we never hear it in full, mathematically pristine form.

The harmonic series, sine tones, white noise, absolute silence: these ideals of sound tantalize the mind's ear. But the world is richer, subtler, more nuanced and more complex than we imagine. Almost nothing we hear conforms to the patterns we construct.

When I was younger, I used to say that I understood my work as a small part of creating a new culture, a culture I might never live to see. This may have been the self-conscious posture of a young artist. But now, many years later, I find myself thinking the same thing, with a more mature sense of what this might actually mean.

For one thing, I now have a more vivid sense of my own mortality. For another, the social optimism of the late sixties and early seventies has long since faded. Over the years it's become clear that politics will not solve the problems of the world anytime soon. Although humanity has made some collective progress, many of our problems have only gotten worse. And we now face problems such as anthropogenic climate change that were largely unimagined a generation ago.

The self-destructive tendencies of consumer society have accelerated and spread throughout the world. It's always been unjust that the United States consumed so much of the world's resources. Now with China and other highly populated countries aspiring to live the American consumerist dream, the unsustainable nature of that dream is more nakedly exposed than ever before.

Something must change. One way or another, something *will* change. Perhaps *everything* will change. The global economy may collapse. Global war may wreak devastation on a planetary scale. Global warming may bring the end of the human species. Confronted with visions like these, what choice do we have?

We can surrender to the easy self-indulgence of despair. Or we can rededicate ourselves to change. We can choose to believe in the possibility that we human animals can live at peace with one another and in balance with our world. We can imagine a new culture and do everything within our power to help create it.

FEBRUARY 11

In recent weeks I've wondered whether this new world I'm exploring in *The Place* may be leading me away from performance works. That doesn't seem to be the case. For the past few days I've been making sketches for possible pieces for ensembles ranging from a trio to full symphony orchestra.

Clearly I'm just beginning to understand the possibilities inherent in the world of *The Place*. I imagine there will be similar installations in the future, works that stand on their own as self-contained worlds. Those works, too, may contain performances, whether composed, improvised or something in between. Given my love for the sound of acoustic instruments played by human performers, it seems certain I'll continue to compose for them. What excites me most, though, is the sense that the music—whether electronic or acoustic, whether played by computers, people or both—will be unlike anything I've done before.

FEBRUARY 12

Gradually the light returns. The pale cream and blue of the past two sunsets have been exquisite. For ten days now I've been listening to the sounds of *The Place* at different times of day to hear, to *feel* how it resonates with the atmosphere of the particular moment. I have no illusion that the sound world I'm creating has anything like the richness and subtlety of real light in the real world. Still, it does seem to ring true. Whether at noon on a cloudy day or at midnight on a clear night with aurora borealis, the sound of *The Place* evokes the feeling of belonging to each moment in this place.

FEBRUARY 23

Today the composers who enjoy the most support from our society are Hollywood note-spinners and a handful of musical conservatives who enjoy the support of large opera companies and symphony orchestras. The rest of us must piece things together and create our own alternatives.

Electronic technology holds little inherent allure for me. But it may be the means by which I can bring my work to full realization. Mixed media of electronics and acoustic instruments give me the means to create a rich new orchestra. Recordings and the Internet allow me to distribute my work to those who want to hear it. Yet it's unlikely my work will ever speak to a large enough audience to support itself primarily from the Internet or from any other quarter of "the marketplace."

Most of the larger cultural and academic institutions in this country embrace only a narrow range of new music. Smaller institutions with bolder visions usually lack resources. And there's little tradition of individual patronage for experimental work. So the economics remain problematic. Still, as Varése defiantly proclaimed: "The present-day composer refuses to die!"

Many of us composers today feel we are outsiders. But we're "outside" only in relation to a presumed "inside." As artists we've chosen consciously to ground our lives and work outside the monoculture of global commerce. So why should we evaluate our work on the terms of that culture? We're not working within the tradition of European classical music or the mass-market commercialism of pop culture. We're not likely to become rich or famous from our work. This is a blessing. We have something much more valuable than fame and fortune ever could be. We have music. Music is our life. And for us music must always come first.

Our world needs as many diverse biological and cultural ecosystems as it can get. In nature the edges of habitats are among the richest of places. Estuaries, tidal pools, the margins of forests are often where life is most profuse. The fringes of music and culture are equally vibrant and productive places. And composers today are discovering a profusion of rich new musical habitats.

Neither the mausoleums of classical music nor the shopping malls of commercial culture encourage real change. Those of us who believe that music can help change the world must use whatever tools we can get our hands on to envision and create change.

We should always be cognizant of the ways in which the tools we use influence our work. Yet ultimately it is artists ourselves who must be the instruments of change.

MARCH 7

I've been listening to Lassus, his *Prophetiae Sibyllarum* (*Sibylline Prophecies*). What elegant, strangely haunting music this is! Its harmonies are diatonic, but melodically all twelve tones are in play. The effect is vertiginous, with little sense of stability as the lines float and meander through a chromatic tonal space.

It occurs to me that *The Immeasurable Space of Tones* traverses similar territory. The difference is that in Lassus the lines are primary and the harmony arises from them. In my piece what sense of line there is arises from the unfolding of the harmony. Not surprisingly, the Lassus is vocal music and *Immeasurable Space* is instrumental music.

The Lassus is full of little surprises, twists and turns of line resulting in unexpected harmonies. My piece seems to proceed inexorably, but because of its diatonic limitation on the harmonies at any moment, the lines (such as they are) sometimes move to unexpected tones. I'd like to find a way to integrate both these approaches in a music that feels both unpredictable and inevitable.

MARCH 9

Last night Julian and I listened to some of my current work. Afterward he observed: "With this noise work you're doing, you seem to have found a viable alternative to harmony." He may be right.

Rather than building up sonorities from a predetermined set of tones, I'm listening for tone within broadband noise, filtering the noise until I find its essential inner resonance. Whether synthetic noise (as in *The Place Where You Go to Listen*) or the noise of percussion instruments (as in *The Mathematics of Resonant Bodies*), the fundamental approach is the same.

Working on *Mathematics,* I had to resist the temptation to employ pitch shifting to "tune" the recorded instruments according to composed harmonic relationships. But I was determined to find the essential resonance of each instrument and let nature take her course. I'm glad I did. This resulted in rich and complex sonorities that would have been difficult to predetermine. Yet as unusual as these sonorities may be, they have a natural quality to them. They sound *real.* They resonate.

If I had to sum up this new approach in a single word, that word would be "resonance."

MARCH 14

Once again it's the season of light. Each day is noticeably longer than the last. The sun moves higher in the sky and traces a wider arc from sunrise to sunset. I find myself waking up earlier in the morning. In recent years I've felt sadness at the passing of the night. But this year that feeling isn't as strong, perhaps because we had more snowfall and more cold weather. Even so, as I sit up late writing, I light three votive candles on the windowsill to mark the passing of the darkness.

MARCH 17

Jim is back from a winter traveling in Asia. Today we resumed our work on *The Place.* We began by implementing a new amplitude curve that utilizes the position of the sun above and below the horizon to determine the amplitudes of the Day and Night Choirs. Listening to the sounds of midnight and midday on the solstices and equinoxes revealed a wide range of "colors" from winter to summer. This was the desired effect.

Continuing, we tested three new harmonic scrims I've devised for the Aurora Bells. Based on subharmonic primes 2 through 13, 5 through 19 and 11 through 31, these "falling scrims" complement the existing "rising scrims" constructed on the same primes, adding a lovely sense of tonal depth to the sky sounds.

If we'd already been at work for a month or so, this would have been a productive day. The fact that we accomplished so much on our first day is very encouraging.

MARCH 18

Jim and I have had another productive day in the studio. We continued refinements to the Aurora Bells, adding two new scrims (one rising and one falling) based on the primes 17 through 41. Tonight as I listened to the new array in real time, I had a vivid sense that *The Place* is beginning to come to life.

MARCH 19

Encouraged by the success of the new amplitude curve for the harmonic fields and the new scrims for the Aurora Bells, we turned today to a question that's been on my mind for quite some time: how to give each sunrise and sunset its own unique coloration.

No matter what date and time of day, whenever the sun is on the horizon both the Night Choir and the Day Choir will encompass the same range of frequencies, at the same amplitude. So I've decided to vary the audible colors of night and day in the same way the colors of the real sky change from day to day: with cloud cover and visibility.

The bandwidth of the day–night noise sweep will range from two octaves in heaviest overcast to four octaves under cloudless skies. This makes a dramatic difference in the perceived volume and brightness of the noise passing through the scrims and in the harmonic colorations that result. In addition to these variations in the width of the noise sweep, the bandwidth of the individual tones in the scrim will vary from one-tenth of a semitone (in clear conditions) to a full semitone (in heavy overcast). This yields a wide range of timbres, from almost organlike sounds to breathy vocal sounds to tones on the verge of dissolving into noise.

Over the past few days, the sonic palette of *The Place* has expanded dramatically in range and subtlety. Even after the discoveries of *Strange and Sacred Noise* and *The Mathematics of Resonant Bodies,* I couldn't have imagined what a profuse spectrum of colors is hidden in a band of noise.

MARCH 20

Immersed again in the world of *The Place,* both Jim and I are experiencing something that happened to us last fall. These sounds are seeping deep into our minds. After long hours extracting discreet tones out of broad bands of colored noise, our ears have become highly sensitive to hearing tone within the breath of the world around us. Leaving the studio this evening, we stop dead in our tracks. The wind is flowing over the ridge and down through the forest. It sounds to both of us like a choir of angels singing. This is a strange and wonderful new way of listening.

MARCH 21

Today is the spring equinox, one of the four major solar events of the year. Living in the North all these years, I've become acutely attuned to the annual cycles of light and darkness. Working on *The Place* has deepened this awareness. The solstices and equinoxes are the benchmarks I return to again and again to hear the sounds of night and day at different times of the year.

Although the sunlight now has the quality of spring, the weather these past two days has been howling winter. Strong winds have blown continuously, making the single-digit Fahrenheit temperatures feel much colder. Last night in the wee hours I sat sipping single malt, watching snow devils dance in the bright moonlight. This morning I woke to find that the field in front of the house is a completely new snow-sculpted landscape.

Jim and I continue today, adding two more harmonic scrims to the electromagnetic bells, fine-tuning the bandwidths of the Day and Night Choirs and adjusting the overall amplitude levels of *The Place*. We're working quickly, but precisely. In a matter of days our palette of timbres and harmonies has been transformed from black and white to vibrant colors. It will take a little time for my ears to adjust, to learn to hear these new colors more clearly. As I do, it's likely I'll want to make further refinements to them. For the moment, though, I'm savoring the excitement of discovery.

MARCH 23

Today we hung the moon. A single band of filtered noise, it floats through the sonic sky of *The Place* following the path of the real moon in the real sky.

Just as the light of the moon can transform the night sky, the effect of this aural moon on the atmosphere of the sonic night is magical. This single voice adds remarkable depth and coloration to the breath of the night sounds. It provides a long melodic line that moves at its own speed, independent of the harmonic fields through which it floats. I'm delighted. Appropriately enough, tomorrow night will be the full moon.

Continuing my search for more melodic elements, Jim and I re-tune the Aurora Bells, separating the magnetometer stations into five discreet bands of sound. This requires voicing them in relation to the aural horizon of *The Place* (101.73 Hz) rather than from the solar zenith (426.53 Hz). As with tuning the moon, the effect of this change on the tonal space is dramatic.

Our work right now involves a kind of sonic chemistry, or alchemy. Introducing new elements can set off potent reactions with the other sounds, transforming them into something entirely new and surprising. The rate at which we're making these discoveries is dizzying and exhilarating. At one moment this afternoon I found myself literally bouncing up and down with excitement at the sound.

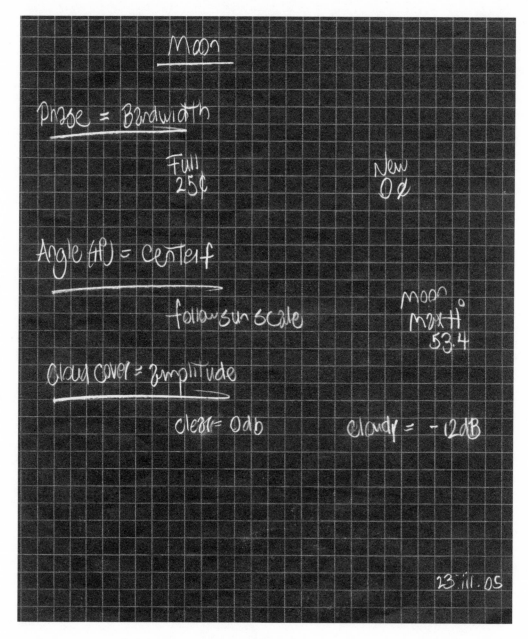

Figure 3.
Notes on tuning the moon. (Composer's sketchbook.)

MARCH 24

The Place is an enormous project, and there's still much to be done before we'll be ready to install it in the space. But I feel the piece now has a life of its own. If I were to disappear tomorrow, there are enough people with knowledge of and faith in the project that I'm confident it could be completed in something like the form I imagine. I have no intention of departing anytime soon. Still, this is a comforting thought.

MARCH 25

Back in the studio today, Jim and I audition further refinements I've written out for the Aurora Bells. The result is ravishing. The sense of line and breath, harmony, register and counterpoint sounds like a dream of Renaissance choral music sung by a choir of other-worldly voices.

MARCH 27

This work is a rich interplay of intuition, logic and perception. I imagine a sound. We employ logic to create it, mapping geophysical data into digital sound synthesis. We listen to the sound. Intuition suggests how it might be refined. We refine our logic and listen again. This process continues until the intuition of the ear tells me the sound is complete.

Beauty is the reflection of our own perceptions in the world around us. Truth is that which we recognize as whole, with or without us.

MARCH 28

Tonight my good friend Leif Thompson comes up for a studio visit. For the past couple of years Leif has been living out in Dillingham, where he practices medicine at the tribal health center. So it's been a while since we've shared a long evening of music, conversation and whisky tasting. After listening to *The Place* for a few minutes, Leif offers a typically astute observation.

"It all goes back to the Aeolian harp, doesn't it? Ever since your wedding trip—what was it, sixteen years ago?—you've been trying to create something like this. Now you've finally done it."

I was floored. This hadn't occurred to me. But Leif is absolutely right. *The Place* did originate during those endless Arctic days listening to the tundra wind through the strings of the Aeolian harp.

He continued: "But this instrument you've created goes far beyond the wind harp. The music of this instrument doesn't come from the wind. It comes from the earth, the sun, the moon and the aurora. This instrument lets us hear things we haven't heard before."

Over the years I've relied on Leif for this kind of insight. Whether challenging or confirming my own understanding of my work, he has an uncanny ability to go right to the essence of the work at hand. Like the best performing musicians, Leif often understands the heart of my work better than I do.

MARCH 30

Jim and I continue to tune the Aurora Bells. Last night I spent some time listening to the new sky we've mapped. It's very beautiful. But there were moments when I felt the harmonies were a little too dense, on the verge of melting into rough-edged timbres. So I revisited the bell scrims and simplified the harmonies.

The previous scrims encompassed harmonics and subharmonics based on all prime numbers up to 61. This produced a field of thirty-six different tones. The new scrims extend only to harmonics and subharmonics based on 31, yielding a total field of twenty-two tones. This afternoon Jim and I tune up the new scrims and listen. The resulting sonorities are exactly what I'd hoped for. At times they hover on the cusp between harmony and timbre. Yet regardless of how dense or sparse the musical texture may be, they always have a wonderful clarity and sense of depth.

APRIL 1

Once again I find myself sculpting with noise. All last summer as wildfires raged, I spent long days in the studio sketching a series of *Noise Veils*. Tonight I've returned to that work.

Last summer I was working with software that required me to write long lines of numbers, make recordings of the noise, and then edit those recordings. If I wanted to try a different length of time or a different bandwidth for the noise, I had to start from scratch.

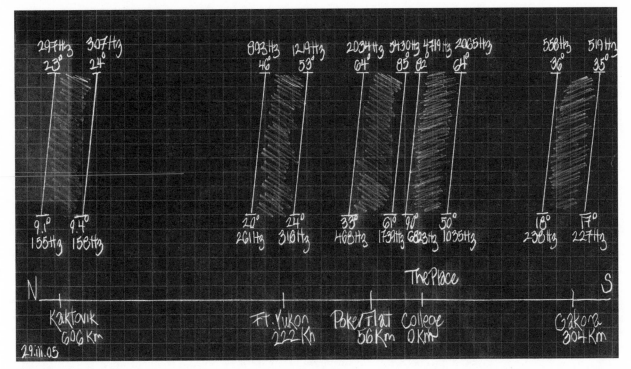

Figure 4.
Map of the aurora in The Place.
(Composer's sketchbook.)

Now Jim has programmed the first version of a "noise factory" that allows me to vary these mists of sound in real time. I've dreamed about these pieces for almost a year, and it's thrilling to have the capability of working with them in this way. So thrilling that it's difficult for me to leave the studio tonight.

My fascination with noise as the basis for music goes back to *Strange and Sacred Noise*. That work led me to discover the world of resonance hidden within broadband noise. I explored this further in *The Mathematics of Resonant Bodies*. Last summer my first sketches of the *Noise Veils* led me to the idea of a year-long sweep of noise controlled by the sun, which became the central element of *The Place Where You Go to Listen*.

Now I'm working with noise in its raw, most primal state. Eventually I may filter the noise through harmonic scrims like those in *The Place*. But for the moment I'm working with pure pink noise. My objective is to leave it as untouched as possible, to let it color time and space with atmospheres as simple and as beautiful as they can be.

APRIL 3

Spring is taking its time in coming. It's not cold the way I remember from years ago. Still, it's not warm. And the snow continues to fall. The sun is back. But although the days are long, the night has not yet disappeared. The light is soft and steely at once, in diffuse slate blues and grays. The air is filled with expectancy and peace.

Yesterday and today I've continued to work with the *Noise Veils*. At first I'd imagined these sweeps of noise would move in three ways: up, down and in both directions at once. It didn't take long for me to rule out the bidirectional forms. I'm seeking evenly distributed atmospheres, and the crossing lines commanded too much attention as they converged into unisons and diverged again. Not long after making this decision, I realized that I preferred the falling lines to the rising ones. This is a little surprising, since most of the lines in my music from the time *Dream In White On White* through *for Lou Harrison* rise endlessly. But there's something about these long strands of noise that sounds right coming down, like cascades of water, like snow falling.

I've been working with five-, six- and seven-voice choirs of noise that begin above the range of human hearing and fall slowly over ten octaves to dissolve into subaudible rumbles. Last night I began filtering the sweeps through banks of band-pass filters (my harmonic scrims) tuned in five-tone and seven-tone equal temperament. Today I've continued, adding five-, six- and seven-tone scrims tuned to prime-number subharmonics. This changes everything, as the noise clarifies into tone.

I'm sketching at the computer the way I've sketched at the piano in the past: considering each sound carefully, listening and following my ear and my intuition. There's a wonderful blurring of pitch and timbre in these sounds. As new tones emerge together, widely separated in pitch space, it's sometimes difficult to tell how many tones are in the air. They tend to meld together into a single rich sonority, the higher pitches sounding like inharmonic overtones of the lower tones.

The equal-tempered modes and the "just" (whole number) modes not only sound different harmonically, they also produce different rhythms. The rate at which the noise falls is constant. Because the tempered tones are evenly spaced in pitch, the noise passes through them at regular intervals of time. In the "just" modes each interval is a different size, which produces more varied rhythms. After years of dreaming about it, I may at last be on the verge of realizing a truly unified field of sound, in which both pitch and rhythm are determined by the whole numbers of the harmonic series.

As with most of my music, I imagine these veils will move in multiple layers of time, with two or four choirs sounding simultaneously at different tempos. Since the tempos are quite slow, the multilayered composite will repeat on very long cycles—many hours, or even days, long. As with *The Place,* this feels less like traditional composition and more like creating an environment and setting forces in motion within it.

APRIL 4

It's 1:00 A.M. I've just returned to the house following another long, productive day in the studio. The night is clear. The boreal owl is calling. On the way up the hill I stopped and watched, breathless, as an unusually active display of aurora whirled directly overhead. The candles are lighted. The whisky is delicious. Life is very good.

APRIL 6

Things are moving so quickly in the studio that it's difficult for me to keep track of all the developments. In the past two days the *Noise Veils* have continued to evolve. For one thing, they're now simply *Veils*. Although at some time in the future I may complete some composed entirely of noise, at the moment I'm drawn to the sound of the noise strands filtered through the harmonic scrims. The movement of the noise sweeps is still clearly audible, but the sounds are more discreet tones than continuous sliding tones. With the noise filtered through the scrims, I find I like the rising forms just as well as the falling. And crossing forms with lines moving in both directions are beautiful, too.

I've continued to devise numerous tuning configurations that Jim and I have produced and auditioned. This reminds me of learning to discern the distinctive qualities of different single-malt whiskies. So I'm starting to call it "mode tasting." I seem to be settling on a palette containing a few favorite modes. But something tells me we haven't finished tasting yet.

APRIL 7

In the past week I've sketched and tested dozens of harmonic scrims. Tonight I listened to all of them and weeded them down to four sets of four scrims that sound best to me. Each set defines a spectrum of gradations within a certain harmonic color. The effect of moving from one to the next is like adding or subtracting stops to the sound of a pipe organ. This will be my basic tonal palette for the first *Veils*.

APRIL 9

My search for unified fields of sound continues. In the *Veils* I've decided to use tempo relationships based on the same prime numbers as my harmonic fields. In theory this means that, sped up fast enough, the tempo relationships between the layers of sound would produce the same pitches as the sound itself.

Since I'm working with primes up to 31, this will make for close tempo relationships and very long periods for all four layers of the field to complete a cycle. For long-term installations, long periods will be fine. And generally I'm discovering that the slower the tempo, the better I like the sound of the *Veils*. But can I make scale models of the *Veils*? Is it possible to create complete shorter pieces that suggest much longer sweeps of time?

APRIL 12

From my first experiments, I'm not yet certain whether the tempo relationships of the *Veils* can remain fixed while the time scales of the pieces are shortened. But I'm also considering another kind of scale: the scale of the pitch space. At some point I imagine I'll want to make performance pieces drawn from the *Veils*. The pitch range of most instruments and ensembles is smaller than the ten octaves of the electronic *Veils*. And the range of the human voice is even more limited. So I'll need to find ways to suggest expansive tone fields in a relatively compressed tonal space.

APRIL 13

For decades I've composed in conventional musical notation. And the limitations of that toolkit have become transparent to me. Today I got an amusing demonstration of just *how* transparent. Jim and I are designing a tempo calculator that will allow me to determine the length of each layer of *Veils* in any relationship of tempos, at any total duration. We talk through the steps necessary to make this possible: Find the least common multiple of the tempos. Multiply that number by the desired time scale (for instance, sixty for one minute). Then divide the result by each of the tempos.

After discussing this several times, Jim looks at me and says: "Wait a minute. Couldn't we just decide how long we want the whole thing to be and then divide that by each of the tempos?"

Of course we could! In conventional music notation, everything has to fit within the bar lines and the tempo markings of the metronome. But with the computer we can produce virtually any fraction of any duration with a sounding accuracy that exceeds the limits of our perceptions. We're not working with notes here. We're working with *time!*

I always try to be aware of how the tools I use influence the work that I do. But this little epiphany is a reminder of just how deep and invisible those influences can sometimes be.

APRIL 14

This afternoon Jim and I listened to a scale model of a five-tone rising *Veil*. This was the first time Jim had listened to this work for a sustained period, so I was very interested to hear his observations. He was very excited about what he heard.

"I like the strong sense of unfolding as the voices enter and rise. You keep thinking: 'This is as big as it's going to get.' Then it gets bigger." Jim said he loved the sense of immersion in the middle and the way the voices drift upward out of our hearing at the end.

I asked him about the timbres. "Are they too synthetic? Too metallic?"

Jim replied: "I like them." Then he paused. "I'm not sure this is music," he said. "But I think this is right where you want to be!"

I agree.

APRIL 15

As I've listened these past few days, I've come to understand how the number of voices in the choirs determines the melodic and contrapuntal density of the music. The density of the harmonies is determined by the bandwidth of the noise sweeps. And the bandwidth of the individual scrim tones determines the timbres of the sounds.

Last night I made two "scale models" of two *Veils*. Their overall shapes and colors are pleasing, and the atmospheres they create are pretty enough. But at thirty-five minutes for a complete cycle they feel a little too busy, too compressed in time and tonal space . . . a little too much like music! Perhaps they'll be more satisfying if I double the scale.

It also occurs to me to try fewer voices per choir. This may thin out the contrapuntal texture enough to create more of the sense of clarity, space, evolving timbres and textures that I like so much in the full-scale *Veils*. Yet I find myself preferring real-time excerpts rather than compressed and accelerated versions of the real thing.

Writing about *In the White Silence* and *The Light That Fills the World,* Kyle Gann has observed that *White Silence* is a metaphor of eternity, while *The Light* is composed of spun-off shards of eternity itself. Kyle expresses his preference for the metaphor. But working on these *Veils*, it seems I may be better off presenting the shards.

APRIL 17

Back in the studio tonight, I audition different time scales for the *Veils*, ranging from thirty-five minutes to one year. At such widely varied speeds, the same sounds require very different treatment. It seems that the slower the speed the denser the fields want to be, both in the number of voices and in the bandwidth of the single tones. But regardless of the speed, there's a delicate balance between saturating the tonal space and maintaining a degree of clarity and transparency to the textures.

After playing with the time scale for a while, I return to harmony—revisiting the harmonic scrims I'd set aside last week and restoring a few of them to my active palette. No doubt I'll continue to experiment with new scrims, but I imagine I could work for a long time with the colors I now have.

For years I've imagined that eventually I would work with unabashedly synthetic sounds. Now that time has come, and I find myself a bit unnerved by the hard-edged brightness and restive rumbles of these sonorities. At the same time I find them mesmerizing.

I take pride in my orchestrations, and I'm flattered when listeners remark on the instrumental colors of my music. Some listeners may dislike the sounds of the *Veils*. But these sounds seem inseparable from the musical conception of these pieces.

I'm remembering Dan Flavin. The clearly synthetic sounds, limited harmonic palette and strict but endlessly variable format of the *Veils* reminds me a little of his fluorescent light sculptures. But the moods and atmospheres of these new works, especially *The Place,* seem to have even more affinity with the light art of James Turrell. Many of Turrell's works are centered on a site-specific sensing space, which the artist describes as "a space that responds to a space outside with a logic or consciousness formed by its look into that space." Turrell's works are often shaped by the angles of the sun, encompassing both a day aspect and a night aspect. So is *The Place.*

APRIL 18

Morton Feldman, who is known for the length of his late works, observed that after about ninety minutes the composer is no longer in control of a piece of music. Beyond that, Feldman said, it's no longer a question of composition. It becomes a matter of scale. I'm beginning to understand this from firsthand experience.

Until now my longest single-movement works have been sixty to eighty minutes. Now I'm working on a much larger scale. The *Veils* may last for hours. And *The Place* will never repeat. Once I've designed these works and set them in motion, the best I can do is to listen as much as I can and be sure I like what I hear at any moment. If I don't like what I hear, I can make adjustments. But there's no way I can monitor everything that may occur in *The Place*. Rather than painting a picture or writing a novel, the process will be more like tending a garden. Even more, it's like taking a journey. I can decide more or less where I want to go. I can make preparations for the journey. But exactly how things unfold is ultimately beyond my control.

APRIL 19

We woke this morning to several inches of fresh snow. It continued to fall all day. Not wet, soggy spring snow, but large, dry feathery flakes.

Over the past week or so the spring melt had tentatively begun. This will set it back. I'm not disappointed. I'm not quite ready to emerge from the introspective mind of winter into the extroversion of spring. And after several successive spring thaws that arrived much too early, it's reassuring to feel that, now and again, winter can still hold sway here in the North.

As the snow fell I continued work on the *Veils*. Tonight I recorded excerpts of three that I feel confident about sharing with other listeners. Tomorrow I'll turn back to *The Place*. It's very late now. The snow has stopped and the clouds have thinned enough to reveal the three-quarter moon. Through binoculars I watch the flowing clouds, two or three thousand feet above me. And I watch the moon, 239,000 miles beyond. The craters and the mountains of the moon stand out vividly. They remind me of the mountains that surround me here at home. Relative to the immensity of interstellar space, the lunar mountains, like the peaks of the Alaska Range, are very close at hand.

In the woods just outside our back door, the boreal owl is calling again.

APRIL 20

Jim and I are back at work on *The Place*. After a couple of weeks away from it, immersed in work on the *Veils,* I imagined I might want to make some significant changes. Since the Aurora Bells are most closely related to the sound of the *Veils,* we begin there. I'm very pleased with what I hear. We make some minor refinements to the bells, but all of the previous work remains fundamentally unchanged.

The *Veils* are very beautiful. They realize ideas I've had in mind for quite some time. They also open up some intriguing new possibilities. But coming back to *The Place,* I'm struck by how sonically complex and how conceptually radical it is. Working on the *Veils* has been a joy. Part of my pleasure has come from the excitement I always feel beginning new work. At the same time I think I've enjoyed the *Veils* because they feel more familiar. Despite their large scale, creating them is more like the familiar process of composing a piece of music.

Because it never ends, *The Place* requires a different approach to form. Because the micro-topology is unpredictable, it asks for a more fluid sense of texture and continuity. Because the sounds are so complex and so highly dynamic, it demands a new kind of orchestration. But more than new techniques, *The Place* is challenging me to a new relationship with my work as a composer, leading me to a fundamentally different approach to the materials of music and the process of composition.

APRIL 21

The work was difficult today, and there's not much to show for it. We began by making some minor refinements to the sound of the moon. Then we returned to the Day and Night Choirs. For months now we've had these fields tuned in twelve-tone equal temperament. Now with the sound of the justly tuned *Veils* in my ears, I feel it's time to return to these primary sonic elements of *The Place* and attempt to tune them to "just" harmonies as well.

Unfortunately we have a problem. When we tune the Day Choir to whole-number relationships, the harmonics reinforce one another to create a strong buzzing in the fundamental tone. We can make this disappear by using higher prime-number harmonics to create the field. But then a Night Choir constructed from subharmonics of the same numbers clashes with it.

This is a tricky little conundrum. It's a reminder that sound is a force of nature, like wind, water or fire. Working with tones on a grid, such as the notes of the piano, we imagine that we can bend the sounds to our will. But working with sound in a more primal state, we come to understand that we must bend to *it*.

My first allegiance is to my work. My work is my life. It is my faith. It is my refuge. It is the only gift I have to give back to the world. My work originates in solitude. It relies on many others to reach its fullness. But its heart lies in stillness, in that place where I am alone with the beauty and the mystery of the world. All my life I've struggled to find this stillness and to protect the solitude from which it comes. Sometimes I've paid dearly for it. So have others, including those I love most. But I've almost always put the work first. And whenever I haven't, the cost has been even higher.

My solitude will return soon enough. For now, Jim is my daily companion. And he's a godsend. His skills, his sensibility and his personality make him the perfect collaborator for the work at hand. We continue to move forward day by day, sometimes by leaps and bounds.

APRIL 22

Today we plunge right back into our work on the Day and Night Choirs. I'm determined to find a way around the problem of the growling fundamental in the Day Choir. We try narrowing the range of the noise sweep passing through the scrim from four octaves to three octaves. But even at two octaves and less, the growling persists. Instead of pink noise we try using the brighter white noise. The difference is barely perceptible.

We revisit some of the scrims we tuned yesterday. None of them sound any better to us today. We try a few more tuning solutions, all to no avail. So we return to twelve-tone equal temperament. It works very well in all possible light and sky conditions throughout the course of the year. At least for now this seems to be the most practical solution. It accommodates rough approximations of all the tones in the Night and Day Choirs. And its even distribution of pitches produces the kind of undifferentiated sonic atmosphere I'm after.

When we move into the finished space at the museum, I may experiment with altering the tuning in response to the natural resonances of the room. I may also investigate chromatic "just" intonations. And perhaps a fifteen-tone equal temperament would produce an even better approximation of the justly tuned intervals of the Day and Night Choirs. But for now we have an eminently serviceable and harmonious solution. Any alternative will have to be equally responsive to what the sounds want to do.

We finish off today's studio session by revisiting the moon sounds. The full moon now sings through a band of pink noise 408 cents wide (about a Major Third). As the moon wanes, the bandwidth of this noise diminishes until it disappears at the new moon. We listen to the moon sound in a range of phases, seasons and times of day. In every setting it sounds well, adding lovely touches of color to the sounds of night and day.

APRIL 23

Once again my curiosity has gotten the better of me . . . and of poor Jim! Today we spend several hours tuning and listening to several different equal temperaments, from thirteen to twenty-four tones per octave. Most of these turn out to be useless. Some of them produce stunningly unpleasant versions of the Day and Night Choirs. The thirteen- and fourteen-tone temperaments sound well. But ultimately I settle again on twelve-tone equal temperament. Although I'm feeling a bit apologetic about this choice, it makes sense to the ear and it's eminently practical.

We've tuned an eleven-limit version of the fields, based on harmonics and subharmonics with multiples no higher than 11. This eliminates one tone in each field (the thirteenth harmonic and subharmonic), making the sounds a little bit clearer. We've also moved both fields down a Major Second to a fundamental of a tempered G (24.5 Hz). This grounds the fields very close to the fundamental "earth tone" of *The Place* and produces a slightly darker sound overall.

APRIL 24

For the past year or so I've been working on a tricky little riddle: How can I be tough-minded and generous-spirited at once?

The only answer that will matter is one I can live by. And it has to work both ways: The clarity of the mind protects the heart from errors of judgment. The generosity of the heart protects the mind from errors of logic.

APRIL 25

Jim has come up with a method for synthesizing the moon sound using fast Fourier transform (FFT) processing. This has the appeal of consistency, since all the other sounds in *The Place* are produced with FFT. More importantly it produces a richer sound. The sound is so much smoother that I've decided to expand the bandwidth. It mixes beautifully with the sound of the Night and Day Choirs.

This afternoon I heard sandhill cranes calling in the distance. Winter is over.

We're immersed in the music of spring. Around the house, the yellow-shafted flicker is laughing and hammering. Juncos and yellow warblers are trilling. Ruby crowned kinglets are bubbling. In the woods around the studio, the ruffed grouse is drumming. Last night the hermit thrush began singing. This is the music of heaven. Whenever my time may come, I want the song of the hermit thrush to accompany my spirit as it leaves this body.

As I return from the studio, I hear a white-crowned sparrow singing in the dusky field. His phrases have the familiar shape of the white crown. But he doesn't finish them, always leaving off the last two or three notes. Is this a young bird just learning to sing? Is it an old bird trying to remember an old song? Or maybe it's a bird in its prime searching for a new song.

APRIL 28

An ivory-billed woodpecker has been sighted in a Louisiana swamp. When I read this story in today's newspaper I almost wept. The ivorybill is a poignant symbol of the passing of wildness from this human-ravaged world. For decades now the ivorybill has been thought to be extinct. I remember as a young man paddling my canoe through Okefenokee Swamp, imagining the presence of the great bird. The fact that at least one individual may still be alive today pulls my heart between hope and despair. Like the family member of a gravely ill patient, I want to believe in miracles. Yet I fear the worst, for the ivorybill and for us humans.

APRIL 29

The sound scrims have given me a new way of listening. They focus the ear in new ways, inviting us to extend the reach of our hearing deep into the heart of the sound. But my terminology needs refining. Tuned to pass relatively broad, evenly dispersed bands of noise, the metaphor of the scrim seems apt. But tuned more specifically to tones of the harmonic and subharmonic series, these banks of filters are more like stained glass or resonant prisms that reveal the hidden colors of noise.

In addition to purely synthetic works such as *The Place* and the *Veils,* I can imagine listening to the ambient sounds of place through a wide range of scrims and prisms. And just as Turrell's sky spaces modulate ambient light with artificial light, I can imagine natural sounds mixed with and modulated by synthetic sounds. The possibilities of this new medium are vast.

MAY 11

The winter snow pack was good this year. But spring has come early again. Now in the second week of May the snow has melted. Already clouds of dust rise as I walk along the dirt road between the house and the studio.

Several of the wildfires that ravaged Alaska last summer have over-wintered. Burrowing into tree roots and tundra vegetation, they smoldered underneath the snow through the deep cold. This past week they've flared up again. Around town people are already voicing their dread at the prospect of another fire season like last summer.

Last night we had scattered showers. This morning the sky has cleared. The dust along the road has settled. I walk to the studio hoping for the best.

MAY 12

These past few days Jim and I have been listening to the earth. We've been fine-tuning the sounds of the Earth Drums and experimenting with a wider range of seismic data. Just as it has five sets of bells played by geomagnetic data from five magnetometer stations, *The Place* will have five sets of drums played by data from five seismic stations.

Listening to both the drums and the bells, I'm hearing (of all things!) counterpoint. Maybe this has something to do with my recent listening to Ockeghem and Lassus. I've always been more interested in harmony, orchestration and texture. But when I have worked contrapuntally, I've tended to favor four-part textures. So why am I now gravitating to five voices? Maybe it's because I have less control over the details of this new music. I recall

Figure 5.
Notes on tuning
the Earth Drums.
(Composer's
sketchbook.)

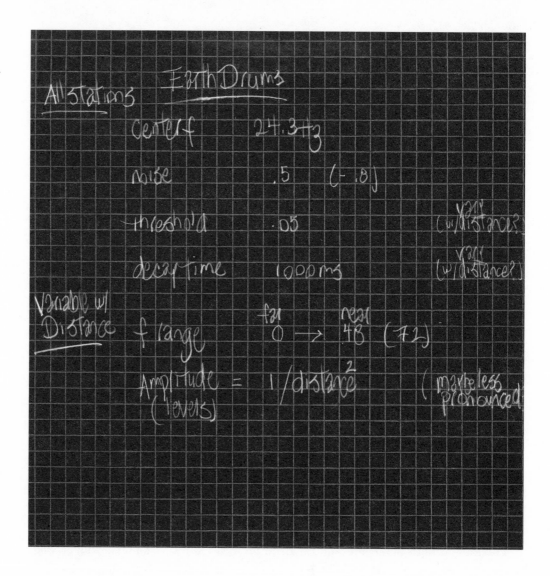

something that John Cage said about the nature of five. With five simultaneous layers, Cage observed, the specific nature of the individual events no longer matters. With five, *everything* becomes interesting.

MAY 14

Here I am again. In the dusk of this spring evening I sit at the piano. The studio windows are open. The hermit thrush is singing. The quiet chords I'm playing meld with the song of the bird. The cycles of my life return me again to the place where I began.

I'm working on a new piece . . . a piece I began in April 1973. I was twenty years old, a student at Cal Arts. I hadn't begun studying bird songs. I hadn't yet traveled to Alaska. But there's something in this little score that, all these years later, I still recognize as my own— as if it had been written yesterday. If I actually finish the piece, the instrumentation will change. The notation will change. The details of texture and the specific sequence of events will change. But the overall sound, the feeling of this music will remain the same.

MAY 15

Today is one of those days when outside feels like inside. The clouds are low. The air is heavy and still. The new green of the woods seems even more intense against the gray-white sky. Rain fell on and off through the night. In years past I might have complained about May rain, fretting that it might mean a cold, wet summer lay ahead. But this year I'm thrilled to see it, and I'm hoping for more. Anything would be better than last summer's hellish fires and smoke.

MAY 18

Over these past few days, Jim and I have finished the studio phase of our work on *The Place*. All the elements are in place, running well on the computer and sounding beautifully. Now Jim will take a couple of months off to visit family and friends. I will move into the next phase of the project, finishing construction and preparing to install the work into the space. I look forward to this with a mixture of excitement and anticipation about the challenges that lie ahead as the imaginary and physical worlds come together.

A Composer's Journal Part III—In *The Place*

We're back on the tundra. Once again Cindy and I have met Gordon at our favorite spot in the Alaska Range for our annual spring camp. Twenty-five years ago we used to come here in the second week of June. Now we come the last week of May. With the changing climate, the season has shifted a full two weeks.

This is the cusp of winter and spring. There's still snow all around. But the tundra is emerging, in burnt brown with patches of new green here and there. A few migratory birds are already present. Others are arriving daily.

Our camp is perched on a high bluff on the edge of a massive river valley. From this lookout we like to pass the long evenings watching the weather, the light and the animals.

A pair of long-tailed jaegers circles endlessly in search of food. A large cow moose with unusual coloration—dark chocolate belly and legs, light blonde head and back—browses nearby with her newborn calf. In the willow and alder flats far below, half a dozen caribou are grazing. The small lakes scattered all around dissolve gradually from rich violet into smoldering gold. To the south the sky is intense lavender. To the north it glowers deep gray-blue.

In the failing light we hear a hollow buzzing that crescendos into a growl. Three tundra swans fly right past us, winging into the dusk.

MAY 28

This day is filled with birds. On a long hike we encounter oldsquaw ducks, rock ptarmigan, golden plovers, a marsh hawk, swallows and sparrows everywhere.

The wind blows. The sun shines. It rains a little. Then the sun shines again. This is spring on the tundra. After the sustained intensity of work in recent months, I want only to lie here in the sun and melt back into the earth.

Today we hike in the opposite direction. Walking up out of the small bowl around a tundra pond, we spot a large bull caribou on the slope just above us. We stop and lie down behind some low willows. We're not sure the caribou detects us. But he shows no sign of concern, going on about his business browsing on lichen while we watch in awe of his beauty and grace.

Cindy takes the lead as Gordon and I linger, continuing our conversation that began more than twenty-five years ago. We talk about music. We talk about the earth. We talk about the precarious future of our human species. We talk about our own lives, and our aspirations to live fully and to give more than we take.

Back in camp the three of us watch fireworks. The sun is brilliant, and there's a stiff breeze out of the north. As wind, light and water interact, the pond just below camp explodes in showers of sparks. Each new gust sets off a new explosion of light. We've never seen as spectacular a display as this sparkling fire on the water. We sit mesmerized until the sun moves lower and the pyrotechnics subside.

All year long I look forward to these spring outings with Cindy and Gordon. Sharing this country I love in the easy company of these two people I love gives me comfort and makes me feel deeply grounded. I hope we'll continue this annual ritual for many years to come. Right now, I wouldn't want to be anyplace else in the company of anyone else.

MAY 30

We are sitting on the last bluff nearest the river. The air around us is still. But we can watch the wind surging like a mountain stream, through the willows and down the draws. And we listen to its voice high above us, singing the contours of the land.

Driving home this evening, Cindy and I spot a pair of magpies along the Tanana River well north of the John Haines homestead. Over the years we've seen magpies in Isabel Pass, in the heart of the Alaska Range. From time to time we've seen them near Delta Junction, north of the mountains. But we've never before seen magpies this far inland, this close to Fairbanks.

JUNE 1

I've returned from the mountains inspired to continue work.

The wild roses are in bloom. Every year they astonish me with their fragrance, color, elegance and delicacy, and with how quickly they bloom and drop their petals. This year they're earlier than I recall them ever being. As I work at the studio this afternoon, an unfamiliar movement catches the corner of my eye. I look up to see a hummingbird hovering outside the window. Just as suddenly as it appears, it darts away. This is the first hummingbird I've ever seen here.

JUNE 3

This evening in the midnight light I completed the little piece that I began thirty-two years ago. What a strange mixture of satisfaction and detachment I feel. I'm happy with the sound of the music and comforted by its continuity across the years. At the same time I recognize that it now has a life of its own, independent of me.

JUNE 10

The Place Where You Go to Listen is a cave. We enter this cave not to withdraw, but to extend our awareness. In *The Place* our attention is focused in two directions at once—on the intersection of the inner and outer worlds, at the cusp of imagination and perception.

JUNE 12

Tonight Cindy and I stop by the pond at the bottom of the hill to watch the swans that are nesting there. As I lift the binoculars, I hear a sound I've never heard in Alaska before: the unmistakable call of a red-winged blackbird. Instantly I'm transported thousands of miles south and decades back in time. For me the call of the redwing is a leitmotiv of Okefenokee Swamp. The distinctive *con-ga-reeeeee-up* was ubiquitous in that watery place that I haunted as a young man. When I came north I thought I'd left it far behind. Now, like so many other things from my former life in that distant place, it has followed me here. The music of the redwing sounds as lithe and bright as ever. But here in this bog near the Arctic Circle, it also has a certain ominous ring.

JUNE 15

As I woke this morning, long shafts of shadow danced and flickered on the bedroom wall. Sunlight filtered between the open treads of the stairway, creating a lovely play of tones and textures, animated by the movement of the leaves on the birch trees just outside the windows.

I thought of the sound of the light in *The Place,* and I wondered: What if the texture of the noise that filters through the harmonic prisms was influenced by the changing speed and direction of the wind? This might produce a lovely fluttering in the sound of the filtered light.

JUNE 17

Every afternoon or evening for the past week we've had thunderstorms. This is not unusual. But day after day the intensity of the thunder and lighting has been more violent than I recall in any of the twenty-nine summers I've lived here.

JULY 5–9

My work inside *The Place* has begun. All this week I've been in the room, hanging insulation and drywall with my friend Matt Kasvinsky and a crew of young volunteers from the Tanana Chiefs Conference. The building is still a hard hat zone, and I have months of labor ahead of me. But my work is now becoming physical in a way I've never experienced before.

The contrast between this construction site and my studio couldn't be greater. In the studio I'm usually alone or working with one other person. The pace is deliberate. The mood is quiet and focused. At the site things are busy and often loud. People come and go, sometimes quickly, and several things are usually happening at the same time. I'm familiar with this mode of working from my experience renovating our house. But this is the first time that my art has taken me into the realm of construction.

JULY 11

Today for the first time I'm alone in *The Place.* I measure the internal dimensions of the space and sketch out the locations of the loudspeakers. All four walls curve and the ceiling slopes, so a simple grid layout isn't possible. Each speaker must be individually located. I begin by finding the center of the room and, from there, the center point of the north and south walls. From these points, everything else will follow.

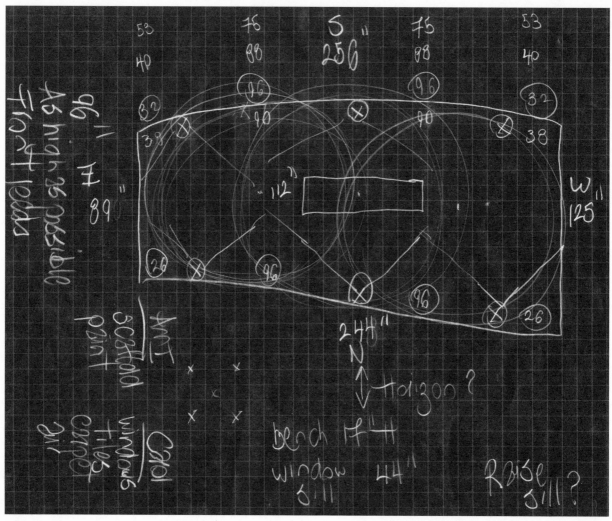

Figure 6.
Sketch of speaker locations.
(Composer's sketchbook.)

I also listen. Some of what I hear isn't good. The air handlers in the room are much too loud. They will need to be modified or turned off. And the compressor for the water fountain just outside the room is also too loud. But we will address these problems in time. The good news is that the acoustics of the room are quite lovely. The gentle curves and hard surfaces impart brightness, surprisingly, without any undesirable reflections.

The room is small. As I stood there just soaking in the ambience, I realized with new clarity that I want to fill the space completely with sound and color. Entering the room should feel like entering a fluid medium. The only object will be a simple bench for seating. Even so, it should be palpably evident that this is not an empty space.

AUGUST 1

Like the sounds of the harmonic fields, the visible colors of *The Place* will be shaped by the rhythms of night and day. But the light will not illustrate the sound. The sound will not accompany the light. Both are part of the same continuum, an enveloping atmosphere that immerses the listener.

AUGUST 5

I've suspended work at the museum. Two days before my subcontractors were to arrive to begin coating the windows, my lighting design was rejected. To compound matters, I've been unable to get any information that might allow me to begin making a plan for mitigating the noise of the air handlers.

Delays in the building construction have already postponed the scheduled opening of *The Place.* This will delay it further. Who knows when I'll be able to resume work in the space. For now, it's back to the studio and back to the drawing board with the lighting design.

SEPTEMBER 20

Although it's still warm for mid-September, the leaves have fallen. Autumn was brief, as always. But winter has yet to arrive. In the past the first snow usually followed soon after the last leaves. Sometimes it even preceded autumn. (I remember one year a heavy snow fell at the beginning of September while most of the leaves were still green. Here and there you can still see birches bent low to the ground from that heavy coating of ice and snow.)

In recent years this new half-season has followed autumn, a strange limbo before winter, with no leaves on the trees and no snow on the ground. The forest is heavy with expectancy, waiting for the cleansing whiteness and the deep cold. I hope there will be lots of both this year.

Tomorrow Cindy and I will fly to New York for a few days. From there, we'll take the train to Boston for the premiere and recording of *for Lou Harrison*.

OCTOBER 1

We're back home from a busy and productive trip. The performance of *for Lou Harrison* by Stephen Drury and his ensemble at the New England Conservatory was beautifully shaped and sounded. This is the most unabashedly lush music I've ever composed. There are passages in which the tonal space is so saturated, the textures so dense that it's almost overwhelming. It's also my most personal work. I think Lou would like it.

I still don't have the answers I need, but we're scheduled to resume work at the museum in another two weeks. I'm determined to make this happen.

OCTOBER 13

John Olson has studied infrasound for over thirty years. Yesterday I called John to ask him a couple of questions. During the course of our conversation he said to me: "You know, John, getting to know you and your work really changed the way I view my own work."

I was touched by John's words, which I received as a generous compliment and the return of a favor.

John was my first scientific collaborator as I began my work on *The Place*. Our early conversations were pivotal in determining my conception of the work. They were also thoroughly delightful. John's mind is extraordinarily quick and flexible. Every bit as creative as any artist, he immediately understood the possibilities for *The Place*. John confirmed some of my intuitions about how science and art might come together in this work. Now it seems this convergence has changed the way each of us understands our work.

NOVEMBER 11

We're back at the museum, with a vengeance. The past two weeks have been a whirlwind of construction. Early each morning Jim and I have driven in to the museum to join our crew of workers installing the sound system in *The Place*. We've been framing, insulating, hanging drywall, running cable and embedding loudspeakers in the walls and ceilings. Each evening I've taken a hot bath, eaten a good meal, packed my lunch for the next day, listened to music and fallen into bed, exhausted.

One day last week, while applying acoustical sealant to the wall frame in the control closet, I cut myself on one of the metal studs. Although it wasn't a bad wound, it bled freely. But I was trying to stay ahead of the rockers, so I kept working. When I turned to receive the next piece of rock, I saw that I'd dripped blood onto my work. The rock was hung and my blood became part of the wall.

For years I've felt that art is a blue-collar profession. Now I'm coming to understand this in a more complete way. Artists are workers. Art is work. And the artist must give himself or herself to the work completely.

While we've been working feverishly in *The Place*, winter has arrived. Several inches of snow now cover the ground. Nighttime temperatures have hovered in the teens below zero. Historically temperatures in early November have been much colder. But what a relief it is that winter, even in this milder form, has found its way home once again.

NOVEMBER 12

Much of my work is grounded in a vision of the world as I dream it might have been or as it might become. But *The Place Where You Go to Listen* is grounded in attention to the magic and the wonder of the world as it is. This may be my first work in which perception takes precedence over poetry.

NOVEMBER 18

Today, for the first time, I heard the music of *The Place* inside the place itself.

Mike Dunham is here to research a magazine article he's writing. His visit coincides with an unusual sky event here in the North: the midday moon. The moon is reaching the end of its 18.6-year cycle, and for the past three days it hasn't set. Like the midnight sun of the Arctic summer, it rolls along the northern horizon and then rises again.

This moment occurred, right around sunset, while Mike and I were listening in *The Place*. Since the sky outside was heavily overcast, we weren't able to see the moon. But inside *The Place* we were able to hear the moon, floating like a full baritone voice just above the aural horizon. This is what *The Place* is all about: extending the reach of our senses, allowing us to hear the inaudible and the invisible.

NOVEMBER 30

Learning to work in this new medium is like learning to do anything. The only real way to do it is to *do* it. I'm intimately involved with every aspect of this project, from hanging insulation and drywall to lighting and programming design. This is a do-it-yourself project. My previous experience designing and renovating our house comes into play. In *The Place,* I'm serving as my own interior designer, my own contractor and my own janitor. For future installations I must have a project manager. But this is a valuable learning experience. I feel a bit like an architect getting a little hands-on experience in the building trades.

DECEMBER 2

Twenty-five years ago today President Jimmy Carter signed the Alaska National Interest Lands Conservation Act (ANILCA) into law. Extending protection to more than 100 million acres of wild lands, this was the largest piece of land conservation legislation in history.

Five years earlier, as an idealistic young man, I came to Alaska to work for passage of ANILCA. I'm proud to have played a very small role in that campaign and grateful it succeeded. In today's political climate it would probably be impossible. But if ANILCA hadn't passed back then, Alaska wouldn't be Alaska today.

In the thirty years since I first arrived here, the human population has increased dramatically. The Trans-Alaska Pipeline was built and, perhaps inevitably, the *Exxon Valdez* oil spill followed. We Alaskans fancy ourselves rugged frontier people. In reality, we're colonists. The economy and the politics of Alaska are controlled by Big Oil. The voters of Alaska (we're not taxpayers, since we don't pay any taxes) have come to regard the annual oil dividend check as an entitlement, while our schools, social services and environmental protections continue to deteriorate. A major natural gas pipeline is now on the horizon. The coastal plain of the Arctic National Wildlife Refuge may be transformed into an oil field. The glaciers are melting and the sea ice is retreating as global warming continues to advance.

In the coming decades, as change sweeps over us like a rising tsunami, the lands protected under ANILCA may be the only places where the unique qualities of Alaska will endure.

DECEMBER 9

Jim has left for the holidays. My friend Paul Lugin and I have spent the past two days cleaning up loose ends and hanging the large door on the "brain closet." This was trickier than we'd expected, but it turned out well. The door reduces considerably the noise from the closet. It has a nice, solid feel. And it closes with a satisfying "ka-THUNK."

DECEMBER 12

In the past two nights, after the museum has closed, I've spent some quiet time in the space looking and thinking about the lighting. I won't really know what I'm dealing with until after the vinyl coating is applied to the windows. Right now there's just too much extraneous light pouring in from outside, and too much of the colored light inside the room is escaping through the windows. But we need to move forward with construction of the alcove to conceal the fiber-optic cables. After the alcove is built and the fiber is installed, we'll apply the vinyl to the windows. Then I'll be able to work with the color harmonies I've sketched out on the computer and learn how they behave in the room.

DECEMBER 13

It's cold, clear and quiet. I spend the afternoon in *The Place*, alone. With long strips of cardboard taped to the wall, ceiling and windowsill, I ponder the proportions of the alcove that Paul and I will construct around the lighting.

In my original conception of *The Place*, color would have emanated from all directions, immersing the listener. In the present space, this just isn't possible. The sound comes from all around. But the wall of windows creates a strong visual sense of front and back to the space. To complicate matters we have only four inches of space to work with between the glass and the edge of the sill.

Maybe in the future I'll have the opportunity to design and build a space from the ground up. For the moment, I need to follow Robert Irwin's dictum: "Play it where it lays, and keep it in play."

DECEMBER 14

As I work in the new museum building day after day, my creative sources are always close at hand. The panorama of the Tanana River Flats and the Alaska Range out the south-facing windows never fails to inspire me. This sweeping view is a touchstone, a reminder that my work comes directly from the earth, the sky, the light and the *feeling* of this very special place on this earth.

DECEMBER 18

Over the past couple of months I've sustained the usual minor cuts, scrapes and bruises that come with construction work. I still bear the mark of a sharp bump on the head I took last month. Today I had my most significant injury to date.

I was in the middle of caulking around the top of the windows. When I descended, absentmindedly, I left my large, heavy caulk gun on top of the ten-foot stepladder. As I moved the ladder toward my next station, the gun flew down and hit me sharply on the nose and glanced off my forehead.

I was stunned. Although the wound was substantial, it didn't require stitches. Once I regained my balance and slowed the bleeding, I climbed back up the ladder and continued my work.

This was an undeniable reminder that I wasn't paying attention. It's difficult but essential to pay attention to the work we're doing, no matter what it may be. And in this kind of work the consequences of inattention can be harsh. I'm lucky this wasn't the case today.

DECEMBER 19

After six full days of work, Paul and I have finished the alcove. I'm delighted. This was a brutal job, but the results make it all worthwhile. It adds a wonderfully indeterminate depth to the plane of the windows. When it's illuminated by the fiber optics, it should float and vibrate with glowing colors.

DECEMBER 22

I'm accustomed to working alone in the quiet of my cabin studio. Working in a busy, noisy public place is very different. People drop by. They ask questions, volunteer observations and offer unsolicited opinions. I'm trying to keep fairly tight wraps on the details of the work. Things may change. And I want to preserve a certain air of mystery and excitement leading up to the opening. We're also very busy. Even so, I welcome the opportunity to talk briefly with visitors about the concept of the piece, its basic elements and the progress of our work. Good ideas can arise from anywhere at any time.

DECEMBER 24

For the past couple of weeks I've suffered from a constant ache and occasional stabbing pains in the left side of my jaw. These past few days it's gotten much worse. This is probably a combination of stress and fatigue, perhaps exacerbated by all the work on ladders and a couple of blows to the head. I'm treating it with hot and cold compresses, lots of ibuprofen and rest. If it doesn't subside soon, I'll have to see a doctor.

The day after tomorrow the painting will begin.

DECEMBER 28

The room is now painted. Tomorrow Paul and I will reinstall the speakers and the lighting. Next week the windows will be coated. The following week the floor will be installed. Then we'll resume work on the lighting design. I'm eager to begin working with the colors. At the same time I'm a little anxious about it. This will be a new experience for me. I'm not sure I'll be able to successfully mix the colors of a rising day field and a falling night field. But the sunset beyond the Alaska Range this afternoon reminds me that this is what I want to attempt.

DECEMBER 29

Today Paul and I reinstalled the wall speakers and mounted the grills and trims over the ceiling speakers. The walls of the space curve, and the ceiling slopes from front to back and side to side. So it was an interesting challenge to place the rectangular grills in a way that looks balanced.

The grills are quite a bit larger than the ceiling speakers. This gave us the option of placing the visual center of a given speaker away from its actual sonic center, hidden behind the grill. Earlier this fall we'd installed the ceiling speakers along a pair of chalk lines running roughly parallel to the curved walls. Because of structural obstructions in the ceiling, two of the ceiling speakers had to be located slightly off their ideal centers, but we kept them as close as possible to the lines.

We begin our work today using notations I'd made from those chalk lines before the room was painted. Then we deviate from the lines according to visual intuition. As with the sound, it's surprising how minute adjustments can make a big difference. We move a grill just a little and suddenly the relationships between speakers seem to "snap" into balance.

With the walls painted and the speakers installed, the room is beginning to look quite elegant.

DECEMBER 30

Today we hung the fiber-optic runs along the sill and the ceiling. We had only four inches of latitude inside the alcove. Still, I spent quite a bit of time considering the precise location of the fiber before we mounted it. Paul is remarkably patient and attentive to detail. Even so, he's accustomed to the fast pace of construction sites. My way of working must seem glacially slow to him. But compared with the usual pace of my work in the studio, our current work at the museum moves along at dizzying speed.

Installing the fiber on the ceiling is very physical work. For two hours, perched atop tall ladders, we twist, turn, push, pull, grunt and curse. The room is warm and we're sweating profusely. But when it's finally over, things are right where we want them to be. The next step is to coat the windows and begin working with the lights.

A hot bath this evening is heaven. I expect to be sore tomorrow.

JANUARY 1, 2006

Paul and I worked steadily through the holidays. At the end of this week I'll resume work on *The Place*. From then on, things aren't likely to let up until after the opening. So the next few days will be the quiet time I've been longing for. I hope to rest, read, write and spend a little time in the studio.

Over the past couple of weeks (at the request of the arts editor and with apologies to W. H. Auden) I've written a little piece that appears in today's edition of the *Anchorage Daily News*. After months immersed in my own world, I've welcomed this invitation to reflect a bit on the current state of the larger world.

JANUARY 5

I've been sketching out a large new instrumental piece for a group in the Netherlands. After months creating electronic music and working at the construction site, it's interesting to return to composing a score in conventional music notation. My progress has been slow. (Composing is hard work, and I'm a bit out of shape.) No matter. It's a joy to be working alone in the studio again. I've made a few runs to the museum to check on things and prepare for the next round of work there. But my mind is largely preoccupied with the details of the new score.

The last few nights my two current projects have blended together in my dreams: I'm working in *The Place*, caulking a seam or measuring something. As I work I'm hearing the sounds of the ensemble piece and thinking to myself: "If I can just get this right, the seam will disappear and the sound will be perfect!"

In some respects *The Place* and the new piece couldn't be more different. Yet they both involve material and they both require construction. In both I want the material to be as strong and simple as it can be. And I want the construction to be so elegant as to become invisible.

JANUARY 6

The relationship between artist and institution is an uneasy one. I've experienced this throughout my life, working with orchestras, theater companies, universities, conservatories,

PRAYER FOR THE NEW YEAR

May we extend to all people and to all life on this earth the dignity and the rights we expect for ourselves.

May we refuse to be ruled by fear.

May we resist all who would impose their beliefs on others by intimidation, coercion or force. May we surrender our own efforts to control the world by these means.

May we forswear violence—violence against one another, violence against the earth.

May we know peace within ourselves. May we embody it in our lives. May we lead by example.

May we say no to fascism masquerading as patriotism.

May we renounce those who would steal our votes and trample our liberties, who would commandeer our government to spy on us, its citizens. May we repudiate them as they turn their deaf ears and blind eyes to torture.

May we demand that our public servants behave as leaders and as elders, not as hand servants of greed. May we, the people, insist that they bear our grandchildren's grandchildren in their hearts and minds with every decision they make. May we turn out all those who betray this trust.

May we conduct our lives as if our neighbors, our future generations, our fellow species and our home, this sacred earth, matter as much as we ourselves do.

May we cherish and defend the precious remnants of our original landscapes.

May we save our dying villages and abandon the sprawling shopping malls.

May we protect the Arctic for centuries to come, not as an industrial park, but as a refuge for the human spirit.

May we relinquish our addiction to burning fossils. May we subdue the wildfires and quell the storms we have created with our lust for speed and power.

May we resolve and act, in our individual and our collective lives, to stem the melting of the polar ice and the rising of the seas.

May we understand that the more life we destroy on this earth, the more likely we are to destroy ourselves.

May we reject the theology of the apocalypse and the politics of death.

May we reaffirm our faith in life—in the rich, miraculous vastness of life, of which we are but a tiny part.

May we stop waiting for the end of the world. May we give up the conceit of escaping to other planets. May we live with the conviction that there must be a future for our descendants here on this earth.

May we acknowledge that we are lost in the darkness, and renew our commitment to the light.

May we eschew joyless possessions and hollow entertainments calculated to make us forget. May we reclaim imagination for ourselves.

May we share stories. May we read and write books. May we make and listen to music. May we dance. May we paint and sculpt and film the shapes and textures and colors of our dreams.

May we look and hope to see. May we listen and hope to hear.

May we celebrate what it means to be human. May we remember that we are also animals.

May we never forget that everything we are and everything we do comes from and returns to the earth. May we recognize that, divorced from what we call "nature," we devolve toward oblivion.

May we live by the knowledge that our own true nature is to be one with Creation.

newspapers and the broadcast media. However, this is my first extended experience working within a museum. Museums need the work of artists. And artists can benefit from the support of museums. But a living artist working within a museum poses substantial challenges for both the museum and the artist.

For months I've felt a passive resistance of *The Place* from some corners of this institution. Recently the resistance has become more active. It's been manifested in countless small ways. Yesterday we reached a moment of truth. The construction engineers object to my plan for treating the windows. At the last minute I was notified that I couldn't apply the vinyl coating to the glass. The coating is an integral element of the lighting design. Without it I can't complete the piece.

The real problem here is not the window coatings. It's *The Place* itself. Unlike a painting or a sculpture, *The Place* is not a discreet object to be exhibited within the larger space of the museum. Physical elements that usually fall under the purview of engineers and facilities managers are integral elements of this work of art. This work *is* the space.

JANUARY 8

This is a challenging experience for everyone involved. Composing in my studio, I have the illusion that all the materials of my work are within my control. Working on a construction site, there's no such illusion. As the new museum building has evolved, I've been forced to adapt to the changing reality of the larger setting. I'm doing my best to play the cards I'm dealt. Ultimately, though, my first obligation is to protect the integrity of the work. This must be the basis for every decision I make and every action I take.

JANUARY 10

Today, after several days in limbo, I received approval to move forward with window coatings. I'm worn down from all the uncertainty and aggravation. But I'm relieved and eager to complete the work.

JANUARY 12

This has been a difficult week. One technical or institutional difficulty has followed another. The building contractor prepared the concrete for installation of the vinyl flooring. But the workers left a thick coating of gray concrete dust all over the freshly painted white walls of the room. The flooring has now been installed. Still, there are problems with the quality of that work. I've been cleared to move forward with the window coating. Yet the engineers remain vocal about their disapproval. And the installation of the coatings can't be rescheduled until the end of next week.

The opening is just over two months away, and I'm now two weeks behind schedule. Physically and emotionally, I'm hanging by a thread. But I can't let down until *The Place* is opened. So I'm doing my best to keep my balance and sustain a steady pace to the finish line.

JANUARY 13

Jim is back. This morning, for the first time in almost six weeks, he walked into the space and looked around, commenting on the changes that have taken place in his absence: the new door on the "brain" closet, the new floor, the painting of the room, the trims on the ceiling speakers. What he didn't notice was the lighting alcove surrounding the windows. Jim is a very observant guy. He's as intimately familiar with *The Place* as anyone. As soon as I pointed it out to him, he remarked on the elegant proportions of the alcove and the way the glass seems to float within it. Yet initially the alcove was invisible to him. I was delighted.

This is one of the most important visual elements of the piece. Designing and building it took more time and energy than any of the construction work we've done in recent weeks. And yet it disappears. This is exactly what it's supposed to do! This is what Jim's programming work does, too. And it's what I aspire to for my work as well. The composer, the programmer, the carpenters, everyone disappears. Only the place itself remains . . . *The Place*, and the solitary listener.

JANUARY 14

In *The Place* I'm working with subtractive color, filtering white light in a way similar to the filtering of white noise that produces the sounds. The colors are produced by passing the light through cyan, magenta and yellow filters. My blank canvas is the wall of glass. The parallels with stained glass windows are inescapable. But the colors of this stained glass are always changing.

Last week the glazier sent me a one-foot-square piece of glass with the coating applied on both sides. This afternoon while we were testing the lights, I placed this square on the sill inside the window alcove. It received the color beautifully, making a smooth, bright field. I was very encouraged. And I'm looking forward to concentrating on the lighting design. But until the windows are coated, there's plenty of other work to be done.

JANUARY 15

It finally caught up with me. I've been sick in bed all weekend. I hope I'll feel well enough to return to work tomorrow morning. It's late, and I need to go to bed. But the stillness is so exquisite. How I love these deep blue winter nights! There's an indescribable sense of expectancy in the air, as if time has frozen and the world is about to begin again. All my life I've been longing to find this feeling in music. Perhaps it has no sound. Or perhaps the sound is so faint that we hear it only as silence. Perhaps the expectancy itself is the very thing for which we're waiting.

JANUARY 16

We're back at it today. Jim is trying to get the geophysical data streaming. After running around town gathering tools and materials, I spend several tedious hours cleaning the heavy gray concrete dust off the walls and the ceiling. This is all part of the work.

JANUARY 22

The days fly by. It's late Sunday night. I've just gotten home after a long weekend in *The Place*. Working with a mom and pop team from Anchorage, we've coated the glass wall inside and out, transforming it into a screen for the fields of light that create the visual atmosphere of the space.

When I'm working alone composing a score, there's the illusion that perfection is at least theoretically possible. But when I work with other people in the real world, such illusions rapidly disappear. I'm not able to do all the work myself, and I can't hold everyone working on the project to the same obsessive standards of perfection that I observe when working alone. The drywall and the flooring have imperfections here and there. The same is true of the window treatments. Still, the overall effect of the work is moving toward wholeness.

I think of powerful paintings I've seen on canvasses that contain patches or other imperfections. Or musical performances with wrong notes that still convey the essence of the music. I have an intellectual understanding of the Japanese concept of wabi-sabi. Now the complex physicality of *The Place* is forcing me to embrace this in a more visceral way. We open in two months, and I must remove my attention from any imaginary ideal of perfection to the larger wholeness of the work.

JANUARY 23

Today is my birthday. I woke up this morning feeling physically every bit of my fifty-three years. But creatively I feel more alive than ever.

I took the morning off and arrived at the museum in time for an early afternoon meeting. The new coatings on the windows are beautiful, inside and out. Just before sunset I got my first look at the lighting inside the room during daylight hours. As I'd feared, too much daylight is getting through and washing out the colors of the artificial lights. It's clear that we're going to have to add another couple of layers of vinyl. This is not unexpected. It will take more time and money. It must be done.

JANUARY 25

Today while Jim is programming, I'm working on the problem of color in the space. It's a bright, sunny day. (The temperature is down around forty below.) And the afternoon sun falls directly on the windows of *The Place*. So the conditions are perfect for troubleshooting.

With just a single layer of frosted vinyl on the inside of the glass, there's a lovely blue-green cast that emanates from the depth of the glass. I'd hate to lose this. But at night our projected colors tend to disappear toward the center of the windows. And during the day (especially a day like today) the color is almost completely overwhelmed.

I've narrowed it down to two possible solutions: we can add a layer of opaque white vinyl, with another layer of frosted on top, or we can add two more layers of frosted.

The first option produces visibly more color in the room. But the higher reflectivity of the opaque produces a flatter surface texture. The second option, two layers of frosted vinyl, produces a softer look. But there is visibly less color reflected into the room. Although I'll miss the softer texture, the more color the better. So I'm inclining toward the first solution.

Even with a total of five layers (two outside and three inside), when the sunlight falls directly on the windows, it backlights the glass and produces shadows from the edges of the building. This was not part of my original conception. But I think I like this interaction of the sun, the architecture and the artificial world of *The Place*.

JANUARY 26

Seeing the glass and the coatings again today, it's very clear to me that the only way to go is with frosted on top of opaque. The atmosphere of the space itself is more important than the surface of the windows. And I want to saturate the space with as much color, as much light as possible.

The Place resides near the edges of perception. Much of its sound and light is subtle. This makes me wonder: How much will a visitor's experience of the work depend on the acuity of her senses? Will younger people, people with keener hearing or more acute color perception have more intense experiences in the space? Or will the experience depend more on the visitor's attitude?

Ultimately, it's impossible to separate sense and sensibility. And I want to believe that curiosity, concentration and receptivity matter at least as much as sheer sensory acumen.

JANUARY 27

Today in a casual conversation a member of the museum staff asked me: "How would you feel about hanging a painting of the aurora inside your space?"

I was stunned. With as much diplomacy as I could muster, I replied that *The Place* is a complete work of art, in and of itself. I've worked very hard to create a self-contained world in which the visitor is invited to experience a new state of listening. To introduce extraneous images or objects into *The Place* would be like placing a representational sculpture in front of a subtle color-field painting to give it more "interest." This would be an act of curatorial vandalism.

I'm not sure the staff member understood. The question betrays a lack of faith in both the art and the museum visitor. And it worries me that well-intentioned people may undermine the integrity of *The Place*. Yet I have to believe that over time the work will reveal itself and that people will come to understand it. What other choice do I have?

JANUARY 28

I'm hoping that in the final weeks before the opening Jim and I will have a few days of quiet in which to fine-tune *The Place* and to savor the experience of being inside of it for extended periods of time. Even more, I look forward to the weeks and months after the opening when I can be alone inside the work in different seasons and at different times of day. More than anything I've done before, *The Place* holds the promise of continuing discovery for me.

FEBRUARY 4

This has been an ordeal. I'm sick. I can't sleep. My budget is worn as thin as my body and my emotions. Things are happening so quickly and on so many levels that I can barely keep track of them all. I simply don't have time or energy to keep up with this journal.

The Place is by far the most ambitious and most difficult project I've ever attempted. Artistically it's more than enough of a challenge. Being responsible for most of the construction and finish work inside the space is far more than I should have taken on. The original plan called for the museum to build a room designed especially for *The Place*. Then I would install the sound and light. Perhaps inevitably, that plan changed. When it did I thought about walking away. However, I didn't. Now I must see it through.

FEBRUARY 5

For years I've espoused the dictum that "wholeness is better than perfection." Now I need to walk my talk. I must embrace this truth in a deeper, more complete way. There are countless flaws in the physical dimensions of *The Place*. But the work is whole. My goal now is to regain my balance and refocus my attention away from the details and onto the whole.

The room is very small. Because I've had to build it myself, I know where every stud, every screw, every wire, every drywall seam is. I see every imperfection, every mistake. These are numerous and they will multiply. But I must look beyond them. I must rediscover the vastness of the original conception. This is my real work. This is what really matters. If I do this, *The Place* will be magic. No matter what the room's physical flaws, the clarity and power of the larger vision will come through.

FEBRUARY 6

After a day off (mostly sleeping) I'm back at work today. Paul is leaving next week for an extended vacation, so I hope we can finish up the construction phase of the project by then. The biggest item on our punch list is the light box and the sign that will conceal the door of the brain closet and welcome visitors to *The Place*. We made a good start on this today. We also need to install the grills over the subwoofer speakers, add trim to the doorway and install the bench.

Looking at our lighting on the freshly coated windows today, I'm pleased. During daylight hours there's still a surprising amount of ambient light coming through from outside the space. But the colors are lovely. And (at least for now) I've decided to embrace the fact that the piece has a day aspect and a night aspect.

Today was a particularly bright and sunny. In the late afternoon, the sunlight flooding through the west windows outside *The Place* was almost blinding. Inside, the sunlight carved a mosaic of yellowish patterns within the light fields on the glass. These forms changed quickly as the sun moved toward the western horizon. Within fifteen minutes they had disappeared. The sunlight adds another tempo layer to the experience. Changing at the rate of a sundial, these shadows are one of the faster elements of *The Place*.

FEBRUARY 7

The work was fast and furious today. Paul and I got the lights mounted and wired, began work on the floor trim, labeled electrical outlets and switches, ordered materials and addressed other miscellaneous details. Tomorrow promises to be even longer and busier. But if all goes well we may finish this phase of the work.

FEBRUARY 9

The light box is finished, the fountain is insulated and the subwoofer trims have been mounted. The sign is at a shop where the text will be applied in vinyl cutout letters. The arrival of the bench is now projected for late next week. Unfortunately, Paul will be gone by then. I should be able to address the few remaining construction items. But I can't imagine how I would have done this project without him.

FEBRUARY 10

In my original conception of *The Place* as an elliptical space with no windows, the visitor would have been surrounded by light. In the space I was given, this simply isn't possible. Even with five layers of vinyl on the glass, there's still more transmission of light through the windows than I'd hoped for. I might have used a layer of black to achieve greater opacity. But I wanted the projected color to reflect like sunlight on snow, so I opted for multiple layers of white on white.

After sunset this afternoon, Jim and I resumed work on tuning the colors. I've been hoping for more richness on both ends of the spectrum, the deep violets and reds. And I'm looking for as much intensity as possible in the output of the lighting instruments. The best approach seems to be to engage only two of the three filters (cyan, magenta and yellow) at any time. Using this method, we quickly arrived at a much more vibrant violet. We improved the red, too, although I'm still looking for greater depth.

The ideal of a unified field of pitch and rhythm has been an aspiration for my work. Now, in *The Place*, it seems this ideal extends to encompass light. I've yet to see the light and hear the sound at the same time. But they are unified in conception and controlled by the same data. And I've experienced enough of both to feel that the combined effect will be strong.

FEBRUARY 13

This morning Jim and I worked in the studio, selecting isolated sound elements (drums, bells, Night Choir, Day Choir and the moon) and three "moments" (winter solstice with aurora, September sunset with full moonrise, and summer solstice with earthquake) for use in interpretative displays and programs at the museum.

This afternoon in the space, we continued work on the lighting. To maximize the output of our illuminators, I've sketched out a plan that never engages more than two filters at a time. The Night Field varies only cyan and magenta. This produces colors from rich purple to brilliant cyan. The Day Field varies only magenta and yellow. This produces colors from rich red to brilliant yellow. When the sun is near the horizon, both fields change hues fairly sharply. This means that (as in the real world) sunrise and sunset will be times of more rapidly changing color.

The next step will be to view the succession of colors throughout the year at high speed. This "fast forward" technique was invaluable in composing the harmonic fields of *The Place*. I imagine it will be equally useful in tuning the color harmonies.

FEBRUARY 24

Things have been moving so quickly with our daily work on *The Place* that I've had neither time nor energy enough to record my thoughts. This morning Cindy and I boarded a jet and flew to San Diego. After the relentless work of these past several months, and before the final push to the opening of *The Place*, it feels good to be getting away for a few days.

FEBRUARY 25

This morning we drove to the desert. Passing through Borrego Springs, we drove east toward the Salton Sea. In late afternoon we stopped and hiked up the rim of a lovely twisting canyon. After years of walking on rough and spongy tundra, walking in the desert seemed easy. Although the plants—ocotillo, cholla, creosote and agave—are different, the spare elegance of the landscape seemed familiar. If ever I were to leave the North, it might be for the desert. I hope we'll be able to return and spend some time camping here. In recent years I've been working with creative capital I stored up long ago. Sitting on the pastel-colored rock listening to a duet between two song sparrows, I felt a strong urge to renew the roots of my work by spending more time out in the land.

FEBRUARY 26

Today was one of the most extraordinary days of my musical life. At noon Rob Esler (a doctoral student at UC San Diego) performed *The Mathematics of Resonant Bodies* inside a lovely box canyon in the Anza Borrego Desert. In late afternoon Steve Schick's ensemble, red fish/blue fish, performed *Strange and Sacred Noise* on a dramatic site overlooking the Carrizo Badlands.

The sections of both works that seemed most effective were those for instruments with broader band noise and longer reverberations: the tam-tams, cymbals, triangles, vibra-phones and orchestra bells. The drums were less powerful. *Noise* in particular relies on reflections, difference tones, interference patterns and psychoacoustic anomalies that arise when enclosed spaces are saturated with high-energy, broadband sound. Outdoors, much of this was lost. Yet something else was gained. As Rob observed, hearing this music in these extraordinary natural settings was an authentic and unforgettable experience.

My music has always been inspired by nature. But virtually all of my work is composed for indoor spaces. After today I'm thinking about composing new works intended from the start to be heard in outdoor spaces.

FEBRUARY 27

This morning Cindy and I drove back to La Jolla for a symposium about the relation-ships between music, art, science and the environment. To open the gathering I read from an essay in progress about *The Place*. The group of scientists and artists responded warmly and raised new questions that I hope to explore in the future. The afternoon continued with talks by other artists and scientists. I was especially engaged by the work of the bio-acoustician John Hildebrand. When, in passing, one of the other artist speakers referred to human beings as an "aberrant species," I winced involuntarily.

We humans engage in destructive, even abhorrent behavior. However, in biological terms, I'm not sure there *is* such a thing as an aberrant species. And in any case it's wildly simplistic to write off our entire species with a rhetorical flourish. Such a pronouncement is a product of the same self-aggrandizing thinking that has led us *Homo sapiens* to the brink of extinction. Besides, it's far too easy. If it's a foregone conclusion that we're an evolutionary aberration, then there's no chance we'll survive and no point in resisting our inevitable decline. We should either do ourselves in or party like there's no tomorrow.

Both false hope and facile despair are self-indulgences we cannot afford. What we need now is clear-eyed, open-eared attention to reality.

Before dinner Rob and his girlfriend, Lisa Tolentino, help me clear the conference room and set up an installation of the *Veils*. After dinner we return to listen. This is the first time I've heard the *Veils* outside my studio. Rather than presenting shorter excerpts in succession, I decide to present all three simultaneously. We've distributed them in three stereo pairs of speakers, deployed in a triangular array around the space. The low frequencies of all three pieces are routed to a single subwoofer in the middle of the room. We carefully balance the amplitude levels so that, sitting in front of one of the pairs, the listener heard primarily a single *Veil*. Sitting in the middle of the room is like floating in the middle of an ocean of sound.

At the end of this long, full day, Steve Schick and I share a dram of whisky at a nearby hotel. I tell Steve how, once again, I've been struck by the strength, clarity and richness of the sounds he produces on his instruments. He avers that this is probably just the result of having played trillions of notes over the years. Following a reflective pause, he adds slyly: "You know, John, I've never really felt that close to sound."

Turning the compliment around, Steve tells me that he admires my relationship to sound. I can't resist quipping: "You know, Steve, I've never really felt that close to *music!*"

We both laugh out loud. This friendship has become very important to me.

FEBRUARY 28

Before leaving La Jolla this afternoon, Cindy and I spend a little time on the beach. She watches birds and explores the rocks nearby. I take off my shoes and socks, stretch out in the sand and listen to the surf. The rich sound and the fresh sea air are wonderfully restorative.

MARCH 6

This evening after Jim leaves, I sit alone inside *The Place*. It's just before sunset and both the Day and Night Choirs are fully present. The half-moon floats high above the horizon. The Aurora Bells begin singing. All the sounds are perfectly balanced in loudness and beautifully distributed around the space. Intermittent rumbles from the Earth Drums set everything else in motion. I bask in the sound, wondering what may happen next.

MARCH 7

Years ago I retreated from politics. At the time it seemed I didn't have the right temperament for a life as a crusader. Now I wonder whether I didn't have enough courage. And do I have enough for a life in art?

Sometimes I feel like a fraud, as though my life doesn't live up to the aspirations of my work. Maybe that doesn't really matter. Maybe at last, as I've been hoping for years, the work has become a world of its own, independent of me and my shortcomings.

Last week, after performing *Noise* and hearing *Mathematics* and the *Veils*, Steve Schick observed: "It's not Alaska. It's not nature. It's not the mathematics or the ideas behind your work that hold our fascination. It's the work itself. It's the world you create in your art."

Steve is a generous and loyal friend. I only hope he's right. But regardless of the ultimate value of my work, Steve's observations speak directly to the inherent value of art. Art has meaning not because it refers to something else. Art matters because it is art.

As I've grown older, I've come to feel that the only thing that may redeem my failings as a person is my work as an artist. I don't take this as a license, an excuse for my character flaws. It feels like an incentive. It raises the stakes. It impels me to give everything to the art, as my best offering to the world.

This faith in art as redemption may be an anachronistic conceit. But I've chosen to stake my life on it.

MARCH 9

The days are flying by. I can't keep track of everything that needs to be done between now and the opening of *The Place*. After months in high gear, I'm finding myself easily distracted and feeling undue anxiety about the small obstacles and annoyances the engineers are raising. But as tired as I am, I must regain my balance and my concentration. More importantly, I must remember that my work is my joy and do my best to be present with it in every moment.

MARCH 11

Today was a brilliant late-winter day. As we approach the equinox, the sound in *The Place*—like the world outside—is growing brighter with each day. Listening in the room this afternoon, I sensed the ecstatic feeling that's so characteristic of this time of year. In fact, it struck me that the sound might be just a little too bright, too much like a church organ. So we adjusted the relationship between the Day and Night Choirs and the solo voices within them. Lowering the choirs ever so slightly brought out the inner voices and animated the overall sound in a rich and satisfying way. For months I've been looking forward to this process of fine-tuning the sound within the space. Now is the time. I intend to savor it.

MARCH 13

Tonight is one of those exquisite northern winter nights. Deep in this blue night, all the colors of visible light are singing, just beyond the reach of my ears. Bathed in the whiteness of this late-winter moon, all the sound of the world falls silent.

I'm a night person. Left to my own devices I tend to work late and sleep late. But working at the museum has required a more diurnal schedule. And the return of the spring light has reminded me that most people will experience *The Place* during daylight hours. As it stands now many of the most beautiful moments, in sound and in light, will occur when no one is in the room. So I'm wondering whether it's possible to balance things—especially the colors— to accommodate this reality, without compromising the integrity of the piece. I'd like to try.

MARCH 14

We began the day by testing my new ideas for the colors. To my delight, they worked like a charm. In the new scheme there will be more red and blue during the regular hours *The*

Place is open to visitors. We also adjusted the balance between the Day and Night Choirs, boosting the Night by 1 dB in relation to the Day. Although they're small, I believe these refinements will make a big difference over time.

MARCH 16

These past two days, Jim and I have been in the zone. Things have been quiet at the museum, and we've burrowed even more deeply into fine-tuning. This morning we extended the range of the Day Field to encompass violet and midnight blue on winter nights. And we made minute adjustments to the colors around sunrise and sunset.

Most of the day we spent tuning sound. The new patch that Jim has devised for spatial location adds a vivid sense of depth to the sounds. We compressed the dynamic range of the Day Choir ever so slightly from winter to summer. Then we added another layer of subtle compression to smooth out the change in loudness as the bandwidths of the Day and Night tones change with visibility. Finally, we adjusted the levels of the subwoofers, lowering them by about 3 dB.

We worked almost continually, listening, critiquing, refining and then listening again. At the end of the day, neither of us could recall everything we had done. But the end result is that *The Place* looks and sounds better than it did yesterday. This was one of the most satisfying days since our studio work last year. Now that the room is finished, I'm relishing this rediscovery of the magic of the piece.

Whenever we want to test changes in the sound and light, we dial up major mileposts in the cycle of the year. Midnight and noon on the solstices and equinoxes are the moments we visit most frequently. For months now I've felt a special fondness for the night side of *The Place*. By contrast the sound of noon on the summer solstice has sometimes struck me as a little harsh. However, tonight, after a long day of tuning, it's the sound and the colors of midsummer daylight that I find lingering in my eyes and ears. This is heartening.

MARCH 18

A recent newspaper article called *The Place* "abstract." I'm not offended. Still it seems to me that *The Place* is much more specific, more concrete than much music. Some people have remarked: "What a great concept." I'm certain this is intended as a compliment, and I take it in that spirit. But *The Place* is not conceptual art. This work is all about experience.

People who want to be spectators may find *The Place* abstract and conceptual. Those who are willing to be active participants may find it much more physical and sensual.

MARCH 19

Kyle Gann is here. After twenty hours of traveling, he arrived last night. Instead of heading directly for bed, he asked to be taken to *The Place*. The Night Choir was full and resonant. The Aurora Bells were singing. The colors were deep blue and red. It felt like an appropriate welcome for an honored guest.

Tonight I dreamed I was in a room. Perhaps it was *the* room. A snow goose was trapped inside, flying about, trying to escape. As I moved into the room to try and help, the bird flew toward the window. It hit the glass hard and started to flail about. I tried to open the window, but the bird caught its neck in the opening and began to strangle. Somehow I managed to grab its feet, throw open the window and release the goose. It flew away, honking loudly.

MARCH 20

Today is the spring equinox. Jim spent the day in the brain closet, polishing up the programming. I spent the day doing an interview, giving a VIP tour, mounting the signs on the door and tying up loose ends before tomorrow's opening.

It was a strange feeling to leave the room tonight, knowing that when we walk in tomorrow it will be open to the public. I'm excited, exhausted and relieved. I'm also feeling a bit disoriented. I'm ready to let go of *The Place*. Yet I'm not sure what my relationship to this work is or should be.

Usually by the time I'm ready to deliver a new score, I've spent weeks, months or even years alone with it in the studio. But it's been months since I've been alone with *The Place* for more than a few minutes at a time. I know there will be more things to be done, polishing details of the work. And I hope that in time I'll come to a more complete understanding of it. But perhaps I'll never understand it. Maybe this is what I've always wanted.

I've long said that I want the music to be larger than I am, richer and more complex than I can fathom, that I want to get lost in the music. Maybe in *The Place* this has finally happened. Maybe this piece can never belong to me. Maybe I must learn to belong to it. I thought I had composed it. But it seems that *it* may have composed *me*.

MARCH 21

The opening was a success. The museum staff rallied, showing great support for the work. A couple of hundred people were present and the lobby was buzzing. I had animated conversations with many people who were eager to share their impressions and questions with me. Some wanted to know all about the science and technology behind the work. Others wanted a purely sensory experience, with little or no explanation of how the piece is created. Both reactions are fine with me. But I find myself feeling more affinity with those who want to maintain a certain margin of unknowing, to allow plenty of space for their own perceptions.

Twice I ducked into the room for a few minutes, listening. People were sitting or standing around the room, most with their eyes closed. I had the feeling that I could almost hear the sound of their listening.

MARCH 24

Snow fell throughout the day. Sitting alone in *The Place* this afternoon, I heard the sound of light falling. After several minutes, a new tone (approximately E, below middle C) emerged suddenly. At first I thought it might be coming from outside the room. But then I realized it was the sun sinking toward the aural horizon. This moment was like seeing—no, *hearing*—the first star appear in the evening sky.

The snow continues to fall tonight slowly, quietly . . .

MARCH 26

This day is saturated in brilliant late-winter sun. The sound inside *The Place* is significantly brighter than yesterday. The harmonic atmosphere is broader, the individual tones more sharply defined. Listening about an hour before sunset, I hear the uppermost tone of the Day Choir slowly dissolve and disappear. After almost three years of working on it, I'm just beginning to learn how to listen to *The Place*.

MARCH 28

Yesterday evening Jim and I were working in *The Place*. Suddenly the drums started pounding. We looked online to see that a 4.7 earthquake had just rocked the Alaska–Yukon border region. It was thrilling to listen as the waves hit each of our seismic stations, one by one. . . . Talk about your rock and roll!

MARCH 30

This journal ends as it began, with questions. I'm still not sure just what *The Place* is. I'm not even sure that it's music. But I'm pretty certain that, as Steve Schick might say: "It's not *nothin'!*"

And something tells me that it's not over yet. Whatever this work may eventually prove to be, *The Place* feels like the start of something I can only begin to imagine, suggesting questions that I've not yet learned to ask.

My life's work has always been haunted by a sense of longing—longing for the natural world as it once was, longing for the human world as it might be. But now my work seems to have led me to a new sense of longing—longing to experience the magical fullness of each moment just as it is, here and now.

Hearing Where We Are

They define space more by sound than sight.
Where we might say,
"Let's see what we can hear,"
they would say,
"Let's hear what we can see."
— *Edmund Carpenter,* Eskimo Realities

The human mind is inseparable from this physical, sensory world in which we live.

Our personalities and our cultures, our individual and our collective identities are fundamentally shaped by the places we inhabit. But in recent times we've lost many of our deepest connections with place. And as we've forgotten where we are, we've also forgotten *who* we are.

We've come to regard the unique sensations of each place as little more than "local color," superficial qualities of a world in which most places are essentially alike. Yet whenever we pay closer attention, we understand the extent to which no two places are alike.

The texture of the light is different in every place. The sounds, the perfumes, the touch of the air are different. These different sensations elicit, even require, different qualities of attention. As the philosopher David Abram observes, each distinctive place provokes its own state of mind, so that as we travel from one place to another we travel from one state of mind to another.

My place is the North. Living here for much of my life, I've come to measure my own work and everything we human animals do against the overwhelming presence of this place. Its influence on my work is profound. And I know that my music would not sound the same if I had made my home somewhere else.

But does the place create the artist? Or does the artist create the place?

Not long ago I completed a piece I'd begun composing thirty-two years earlier. As I played through my original sketches at the piano, I was surprised by how much they sounded like my music today. The open harmonies, spacious textures and sense of suspended time in this music seem to evoke a certain northern atmosphere. Yet it was composed two years before I came to Alaska.

As a young man I came north with the dream that I might discover a new kind of music here, music that might be found only here. I was drawn to the North by the land itself and by my own desire, in my art and my soul, for certain qualities this place represents. This was my romantic vision.

THE ROMANCE OF THE REAL

Reality differs from myth in that it is never completed.
—*José Ortega y Gasset*

The North is filled with romantic visions.

Alaska's status as a popular tourist destination has filled its marketplaces with counterfeit art, kitschy paintings, sculptures, songs, poems and novels grounded in the hackneyed images and clichéd language of the "Last Frontier." Trading in the sentimentality that James Joyce called "unearned emotion," products like these diminish both art and place.

It is the work of the artist to imagine and re-imagine the world, both as it is and as we dream it might be. When old myths and images lose their vital meanings, it is time for new visions.

For decades my music has been grounded in my personal mythology of the North. This set of metaphors, images and ideas has served me well. But as I move toward maturity as an artist, I feel the need to expand and renew the sources of my work. As global climate change continues to accelerate, my romantic vision of the North as a place apart has been challenged in an inescapable way. In this uncertain new world in which we live, I feel an imperative to expand my sense of place to encompass the wholeness of this beautiful, beleaguered planet.

Over the years as my work has matured, its northern qualities have become less obvious and, I hope, more deeply assimilated into the music. I like to say that the music is no longer *about* place, that it has in a sense *become* place. Like a healthy ecosystem, a vibrant work of art must be able to sustain itself. Art may derive energies and substance from the world in which it exists. Yet ultimately a successful work of art must be itself.

Ideas are a means to an end. The end is the art itself. I want my work to have intellectual substance. But the essence of my work is not conceptual. It is sensual and experiential. I don't want the work or the listener's experience of it to be circumscribed by ideas, whether mine or anyone else's. Reality is always larger, richer and more complex than our ideas about it.

The fatal flaw of romantic love is our tendency to use it as a mirror for our own narcissism. We see the beloved person primarily as a reflection of ourselves. When the beloved doesn't meet our romantic expectations, we may feel betrayed. We may try to change the beloved, regardless of whether he or she needs to change or wants to change.

This is also true of a romantic relationship with the world. And the results can be equally heartbreaking. When the reality of the world doesn't measure up to our dreams, we feel we must do something to change the world.

But do we really expect the world to change simply because we want it to?

All my life I've believed in the possibility that one person can change the world, and in the imperative to do so. Yet it's not really the world that we need to change. It's the quality of our *attention* to the world.

At this moment things look bleak for humanity. But at almost any moment in our history things have probably looked bleak. And as desperate as things may be, life is still an unfathomable miracle and the world is still staggeringly beautiful. This is cause for wonder and celebration. Things may get better. Or they may get worse than we can imagine. One way or another, with or without us, the earth will take care of itself. The greatest challenge we face now is our own delusion.

We need a new kind of romanticism, grounded more in sense and less in sensibility. While we believe in and work for change, let us rededicate ourselves to reality. Let us surrender our preconceptions and extend our perceptions. Let us engage directly and passionately with this miraculous world in which we live. Let us cherish the world not for what we want it to be, but for what it actually is.

This requires honesty and a modicum of humility. It requires patience. It requires listening.

THE WORLD IS LISTENING

Listening is a primary mode of understanding. As we listen to the world around us, we come to understand more deeply our place within it. Our listening animates the world. And the world listens back.

Not long ago, visiting in Minnesota, I went walking through the woods and fields. As I entered the woods, a vague aural mist appeared on the edge of my consciousness. I walked deeper into the woods. The sound grew more intense. Approaching a small pond, I recognized it as a spring chorus of frogs, trilling, chirping, groaning, croaking.

As I drew closer to the pond, the music grew louder. Then, all at once, it stopped.

I recalled similar experiences in the past: I would approach a pond. The music would stop. I would continue on my way. This time I resolved to discover what would happen if I stayed.

I found a comfortable spot and sat down. I waited. And I listened. As I listened to the pond, the pond listened to me. In time, when both the pond and I were listening, the music resumed—as boisterous as ever.

At first I reveled in the overall atmosphere, the color and texture of the rich sonic field. Gradually I began to distinguish the low percussive booms of one species from the high peeping of another. In time I could hear the unique voices of individual frogs, sorting out their particular locations, tempos and melodic figurations.

I listened for a long time. Finally I rose and walked away as quietly as I could. To my delight, the music continued.

This music was the result of many individual voices combining to create a larger whole, the manifestation of a larger consciousness: the mind of the pond. Through listening, I had become part of this mind and its music.

As I walked away I wondered: How long would I have to listen before I might learn to sing, to add my own voice to the music of the pond?

Inside *The Place Where You Go to Listen* I hope the listener may experience something like my interlude at the pond. After thirty years of listening in Interior Alaska, I have attempted in *The Place* to integrate my voice into the larger music of this place.

RESONANT SPACE

The physicality of everything has a resonance. . . . The thing is
to maximize the physicality while minimizing the imagery.
— *Robert Irwin*

The Place Where You Go to Listen is a resonating chamber. Inside this space the vibrations of natural forces are transposed and amplified within the reach of our ears. Our attention is focused in two directions at once—on the intersection of the inner and the outer worlds, at the cusp of imagination and perception.

Like the outside world *The Place* is inscrutable. Its sonic atmosphere is complex. Sometimes it seems transparent, other times it seems opaque. It's not always easy to distinguish individual sounds within the enveloping cloud of sound. But if we stay for a while and listen intently, the sound will teach us how to listen.

Like listening to the music of the pond, at first you bask in the overall coloration of *The Place,* the ambience of the present moment. In time your ears begin to focus on individual voices—an alto, a tenor with a distinctive pitch, timbre and location—singing within the larger field.

You begin to hear how the low-frequency sounds of the seismic waves move everything else in the room, creating waves and currents that swirl within the larger ocean of sound. You hear that the currents near the floor are stronger, faster than those higher in the room.

The space is saturated with sound. Your attention drifts or shifts from one sound to another. There is always more detail in the air than you can listen to at once.

At sunset can I distinguish the sound of Night from the sound of Day?

And that particular tone: Is it part of the sunset colors? Or is it the voice of the moon? Fifteen minutes later, you hear that this tone has moved, gliding upward to a different relationship with the larger harmonies of the choirs. Yes. It's the moon.

As night falls the emergence of a single low tone or disappearance of a single high tone becomes a momentous event.

On another morning you hear the sonic colors gradually darken as a snowstorm or a front of thundershowers rolls in from the Alaska Range.

On a sunny afternoon you may wonder: Is that the Aurora Bells I hear? Listening for a while, you hear gentle waves from the ceiling, emerging and receding in and out of the larger sound. Yes. The aurora is singing, even though it's not visible in the sky outside.

The sound and light of *The Place Where You Go to Listen* are synthesized from the mathematics of astronomical cycles and streams of data derived from geophysical phenomena. Understanding some of the science behind *The Place* can focus the listener's attention, encouraging her to become an active participant in a dialog between this artistic world and the larger world beyond. But the primary intention of *The Place* is not to convey information. It is to provoke experience.

Beginning with curiosity and fascination, both art and science demand careful observation. Science translates observation into information. Art transforms observation into sensation and experience.

Information can be measured. It can be verified or refuted. Information can be acquired secondhand, without direct experience. Secondhand information can be useful. But secondhand experience is fraudulent.

Authentic experience requires participation. Participation requires perception. Perception provokes imagination. In turn, imagination creates new experience and new perceptions. The cycle continues.

We enter *The Place Where You Go to Listen* with our thoughts and perceptions from the outside world. Inside we inhabit a new atmosphere, a new world of sound. We stop. We sit down and listen. As we attune ourselves to the mind of *The Place*, we learn to hear its music. Listening carefully, we become part of it.

When we leave *The Place,* the music continues. Perhaps we carry some of it with us, back into the larger world. And perhaps we find ourselves listening in a new way.

THINGS WE HAVEN'T HEARD BEFORE

There is an expectancy to the ears, a kind of patient receptivity that they lend
to the other senses whenever we place ourselves in a mode of listening.
— *David Abram*

Our senses are promiscuous. They want to be one. We have a natural inclination to synesthesia, the union of the senses.

We sometimes regard synesthetic experience as a kind of perceptual confusion between sensations that should be separate. But synesthesia may be a holistic response, a manifestation of the natural inclination of our senses toward unity. Excited by strong sensory stimuli, our minds *want* to hear colors and see sounds.

I experience this most profoundly through my ears. For me hearing is the most synesthetic of the senses, the deepest mode of perception. I'm always looking at things and imagining how they might sound. Conversely, when I hear a sound that moves me, my mind's eye strives to see the shape, the texture and the colors of the sound.

Looking is instantaneous. Listening takes time. The visible stays "out there." The aural "comes inside" of us. The shape of the mountain stands apart. The music of the wind over the mountain enters the body.

Yet in response to strong stimulation of one sense, another sense may resonate sympathetically. We may see a bright display of aurora borealis and hear ethereal hissing or whistling. We may hear the voices of the river rumbling over submerged stones and see deep violet forms in our mind's eye.

Somewhere near this confluence of sound and light is where I experience most fully the unity of the senses. And the more fluid the forms of the sensations, the more potent the experience becomes.

Mapping the auditory world of *The Place Where You Go to Listen* encouraged me to think synesthetically. Considering the colors, forms and movements of the aurora borealis, I imagined how they might sound. Translating the arcs of day and night into sound required a similar leap across senses. Often as I worked I found myself wondering: Can I find the music in the colors of this sunset sky? How might this light, this weather at this particular moment *sound?*

In turn the sound world of *The Place* has exerted a lasting influence on the way I hear. After spending countless hours steeped in these complex, noise-created sounds, I began to hear them everywhere. Now I find that my ears have become more finely attuned not only to the particular timbres and frequencies of *The Place,* but to the breath of the world around me.

A friend of mine is an accomplished oboist. In her thirties she lost her hearing.

Following several years of total deafness, she received cochlear implants. This has allowed her to hear again. Still, as she puts it: "I don't hear the same way. I hear things you haven't heard before."

As she learned to hear again, my friend had to learn to listen. Once, as she struggled to understand someone speaking, her teacher observed her tensing up, leaning forward, straining her neck muscles, working to listen. Her teacher's advice was: "Don't try so hard. Don't think about it. Just listen. . . . Relax into listening."

My friend's listening device allows her to select different modes that emphasize different elements of the sounds around her. Listening in these different modes, she says: "I've come to understand that reality is not as absolute as we sometimes think."

Inside *The Place Where You Go to Listen* she tells me she's heard "intense waves" from the Aurora Bells and felt the low frequencies of the Earth Drums entering her body through the floor. Although the range of her hearing is limited, it may be that my friend has a better idea of how to listen to strange new music than those of us with more acute hearing. She knows how to relax into listening for things we haven't heard before.

The Place Where You Go to Listen undermines the usual expectations we bring to music. In *The Place,* events don't occur in easily recognizable, repeating patterns. There is no narrative, no discourse, no development from beginning to middle to end. The music never ends. We're always hearing the sounds of here and now, unfolding in their own time. This requires more active participation from the listener. It invites a new kind of listening.

John Cage defined "experimental music" as "music the outcome of which cannot be predicted." By Cage's definition *The Place Where You Go to Listen* is by far the most experimental work I've ever composed. I'm still learning to listen to this music. I can describe the daily and seasonal cycles of sound and color. But the specific events in *The Place* are beyond my prediction. What happens next is the outcome of the complex interaction of seismic activity, electromagnetism and weather with the cycles of the sun and the moon. And since the music is open-ended, it has no ultimate outcome.

Consciousness emanates from the intersection of place and perception. Perhaps this is where art and science, romance and reality most closely converge: when we participate in the world with an attitude of curiosity and careful observation, the possibilities are infinite.

And the listening never ends.

An Ecosystem of Sound and Light

They say that she heard things. At Naalagiagvik, "The Place Where You Go to Listen,"
she would sit alone in stillness. The wind across the tundra and the little waves lapping
on the shore told her secrets. Birds passing overhead spoke to her in strange tongues.

Naalagiagvik (The Place Where You Go to Listen) is the Iñupiaq name for a place on Alaska's Arctic coast. Legend has it that a certain woman, sitting quietly in this place, could understand the languages of the birds and hear the unseen voices all around her.

In this spirit I conceived *The Place Where You Go to Listen* as a contemplative space for tuning our ears to the unheard resonances of the earth and sky.

This work is grounded in the geography of Interior Alaska, the sprawling region that extends from the crest of the Alaska Range to the crest of the Brooks Range, encompassing much of the Yukon River basin. The cycles of daylight and darkness in this region are extreme. In summer there is constant light. Winter nights are very long. At this high latitude the magnetosphere is especially active, and the night sky is often filled with vivid displays of aurora borealis. This is also one of the most seismically active regions on earth. The weather, too, is extreme—ranging from deep, clear cold in winter to thunderstorms and raging wildfires in summer. These extremes of light and darkness, geomagnetism, seismic activity and weather are the geophysical foundation of *The Place Where You Go to Listen*.

The architectural setting of *The Place* is a room situated just above the main entry of the Museum of the North at the University of Alaska Fairbanks. A small antechamber houses a display that introduces visitors to the work. The main chamber is approximately ten feet by twenty feet. The ceiling slopes gently from fifteen feet at the northwest corner of the room to thirteen feet at the southeast. Loudspeakers are hidden in the walls and ceiling all around the space. Aside from a single bench in the center, the room is empty of objects. But it is filled with light and sound.

Figure 7.
Satellite mosaic of
Alaska. (Geophysical
Institute of the
University of Alaska.)

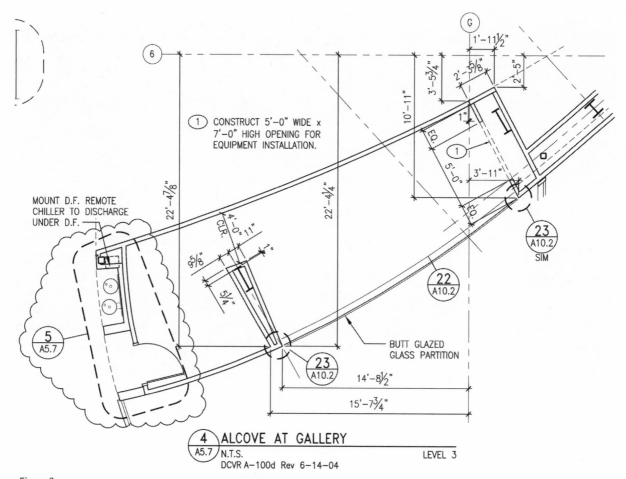

6

G

1'-11½"

3'-5¾"

2'-5⅝"

2'-5"

10'-11"

1"

1

EQ.

5'-0"

3'-11"

CONSTRUCT 5'-0" WIDE x
7'-0" HIGH OPENING FOR
EQUIPMENT INSTALLATION.

22'-4⅛"

22'-4¼"

4'-0"
CLR.

11"

9⅝"

5¼"

MOUNT D.F. REMOTE
CHILLER TO DISCHARGE
UNDER D.F.

23
A10.2
SIM

22
A10.2

BUTT GLAZED
GLASS PARTITION

5
A5.7

23
A10.2

14'-8½"

15'-7¾"

4   ALCOVE AT GALLERY
A5.7   N.T.S.   LEVEL 3
DCVR A-100d Rev 6-14-04

Figure 8.
Floor plan of The Place.
(University of Alaska Museum of the North.)

Plate 1: The colors of The Place—*summer solstice, noon. (Photo © by Roger Topp.)*

Plate 2: The colors of The Place—*winter solstice, noon. (Photo © by Roger Topp.)*

Plate 3: The colors of The Place—*equinox, noon. (Photo © by Roger Topp.)*

Plate 4: Listeners in The Place. *(Photo © by James Barker.)*

MAPPING THE TERRAIN

The Place Where You Go to Listen is a virtual world that resonates sympathetically with the real world. Creating this parallel world involved making maps, exploring and tuning the sonic terrain.

In *The Place,* streams of data tracing natural phenomena (seismic activity, geomagnetism, cloud cover and visibility, the movements of the sun and moon) are transformed into sound through a process that is sometimes called *sonification.* Sonification is not to be confused with *audification,* which is the direct rendering of digital data with inaudible frequencies into the audible range, using re-sampling. (An example of audification would be speeding up an hour of seismological data to play in a second.) Sonification is the process of mapping data with *some other meaning* into sound.

In a sense the data streams constitute numerical maps of the geophysical forces that animate *The Place.* Using these data maps as points of departure, I devised new maps for translating the data into sound. As I listened to the sounds produced by these aural maps, I began to explore the specific features, the detailed topology of the sounding terrain. I then revised my maps to more accurately produce the sounds I heard in my mind's ear, which, inevitably, were influenced by what I heard in the air.

This process of mapping, listening and re-mapping continued until I felt a particular sound had "the ring of truth," resonating in a convincing way with the geophysical force from which it was derived. As the voice of each sounding element emerged, it had to be integrated into the larger ecosystem of *The Place,* in a process I came to think of as "tuning the world."

The data streams that animate *The Place* contain multiple dimensions, reflecting the complexity of the natural forces they describe. So rather than assign single aspects of the data to isolated parameters of the sound (pitch, timbre, intensity, density, duration), I chose to take a more holistic approach.

Both the seismic and electromagnetic data streams include three separate components. But in the sonic cartography of *The Place,* these components are combined. All three components of the electromagnetic data (H, D and Z) are combined to produce a vector that describes the total activity of the earth's magnetic field above a specific location. The seismic data are mapped to produce strong representations of the two types of seismic waves (P waves and S waves) as measured at a specific location.

THE COLORS OF NOISE AND TONE

Nature produces noises not musical sounds.
— *Claude Levi-Strauss*

All the sounds heard in *The Place Where You Go to Listen* are generated from synthetic noise, filtered, tuned and sculpted in a variety of ways. More specifically, the raw sonic material of *The Place* is pink noise synthesized with a random number generator.

Acousticians speak of synthetic noise in a variety of "colors," from white and pink to brown and black. These different colors have different curves of amplitude over frequency.

Just as white light contains all visible wavelengths, "white noise" contains all audible frequencies of sound in even distribution at equal amplitude. "Pink noise" is white noise in which the higher frequencies are "rolled off" (attenuated by 3 dB with each rising octave). This produces noise that we perceive as more equal in loudness throughout the range of hearing, a bit more like the sounds of wind, surf and other broadband noise in nature.

Because the sounds of *The Place* are produced in real time, and because the numbers that produce the pink noise are random, the micro-texture of the sound is always changing. This means that even the most precisely tuned sounds in the work contain minute variations of pitch and amplitude. The phase of the sounds is always random. This gives them a different quality than sounds produced by additive synthesis of tones. To my ear, they sound more alive.

In *The Place*, pink noise is filtered in a variety of ways to create the tones and timbres of the work. Each of the primary elements of the sound world is tuned differently.

- The harmonic fields associated with night and day are tuned in twelve-tone equal temperament.

- The bell sounds associated with the aurora borealis are tuned in prime-number "just" intervals.
- The drum sounds articulated by seismic activity fluctuate continuously within a limited low-frequency range.
- And the sound of the moon is a narrow band of pink noise that floats freely over a wider frequency range.

This varied approach to synthesis and tuning creates an orchestral range of timbres and harmonic colors.

"AN ORCHESTRATION OF UNTOUCHED MATERIAL"

Before *The Place Where You Go to Listen,* most of my music had been created with and for acoustic instruments. But all the sounds in *The Place* are produced with software on a computer. While I was composing with these intangible instruments, Kirk Varnedoe's description of Jackson Pollock's poured paintings became a touchstone for me. I came to think of this work as "an orchestration of untouched material."

I conceived the music of *The Place* for an orchestra of voices, bells and drums. Two choirs of virtual voices create the omnipresent sonic atmosphere, following the arcs and colors of night and day. Low-frequency "drums" rumble in response to seismic data. And high-frequency "bells" ring with disturbances in the earth's magnetic field.

Like an orchestra of acoustic instruments, this new medium offered me many colors and spacious textures. But in addition to composing the music, I had to design and build all the instruments of this orchestra, as well as the physical space in which the music is heard. Yet despite the fact that I was creating a self-contained world, it soon became clear that I could not fully control everything that happens in it.

The Place allowed far less compositional determination of the specific sound events than my previous music. This music has no beginning, no middle and no end. It never repeats itself exactly. Although the instruments don't exist in the physical plane, they are (in a sense) "played" by forces of nature. So my primary work was to conceive of and map this musical world, to design the sounds, to set things in motion and then trust the forces of nature to provide the moment-to-moment music.

In theory *The Place Where You Go to Listen* could be tuned to any location on earth, each of which would have its own sounds and colors. But as I moved deeper into this work, I came to realize just how specific it is to the geography of Interior Alaska. The data streams that animate *The Place* describe natural phenomena that are especially pronounced here. This is my home. And, almost inadvertently, I found myself tuning the world so that this specific location would be the sweetest sounding spot on earth.

THE LIGHT THAT FILLS THE WORLD

And yet there is only
one great thing,
the only thing:

To live to see in huts and on journeys
the great day that dawns,
and the light that fills the world.

—*Inuit song*

The dominant architectural feature of *The Place* is the south wall of the room. A sweep of glass twenty feet wide and ten feet tall in five contiguous panels, this wall is the central visual element of the work. Coated with a frosted white surface and illuminated by fiber-optics mounted above and below, the glass becomes a projection screen for fields of floating color.

A program called SunAngles tracks the position of the sun above or below the horizon and around the points of the compass at any moment in time. These coordinates control the instruments that project colored light onto the coated glass, creating a visual atmosphere of light that changes continuously with the movements of the sun.

Over the course of the seasons, two fields of color traverse the full range of visible light, moving from opposite ends of the spectrum and meeting in the middle. The Day Field moves from red to orange to yellow. The Night Field moves from violet to blue to cyan. (See plate 1.)

The hues of the color fields vary with the height of the sun (H) above or below the horizon. When the sun is above the horizon, the hue of the Day Field moves toward yellow. When the sun is below the horizon, the hue of the Day Field moves toward red.

The Night Field moves in contrary motion to the Day Field. When the sun is above the horizon, the hue of the Night Field moves toward cyan. When the sun is below the horizon, the hue of the Night Field moves toward violet. (See plate 2.)

The colors in *The Place* are always changing. But because this change occurs in real time in concert with the movements of the sun, there may seem to be no perceptible change from moment to moment. Yet within several hours, change is pronounced. Over days and months, from winter to summer, the change becomes increasingly dramatic. As in the real world, sunrise and sunset are the times when the colors of *The Place* change most quickly.

Like the sounds, the colors are created by a subtractive process. White light is filtered through cyan, magenta and yellow (CMY) filters. To maximize the intensity of the light, no more than two filters are engaged at a time. (See plate 3.)

Table 1 shows the hues (in CMY values, from 0 to 255) for the Day Field at significant moments in the year. Table 2 shows the hues for the Night Field at the same moments.

TABLE 1.

Day, Time	Hue	CMY Values
Winter Solstice, Midnight	Midnight Blue	230/255/0
Equinox, Midnight	Deep Red	0/255/255
Summer Solstice, Midnight	Orange	0/210/255
Winter Solstice, Noon	Peach	0/155/255
Equinox, Noon	Yellow	0/70/255
Summer Solstice, Noon	Maximum Yellow	0/0/255

TABLE 2.

Day, Time	Hue	CMY Values
Winter Solstice, Midnight	Violet	210/255/0
Equinox, Midnight	Violet	210/255/0
Summer Solstice, Midnight	Royal Blue	220/255/0
Winter Solstice, Noon	Azure	255/200/0
Equinox, Noon	Pale Blue	255/75/0
Summer Solstice, Noon	Maximum Cyan	255/0/0

The wall of glass is coated with five layers of white vinyl and five coats of white paint. Even so, during daylight hours a perceptible amount of ambient light from outside the room penetrates the world of *The Place*. This enhances the contrast between the day and night aspects of the work.

As the sun rises and sets each day, the lighting design passes through the same hues. But the sunlight and the weather outside the space modulate the colors inside. The angle of the sun and the rate of change vary from day to day. These changes and changes in sky conditions make each sunrise and sunset in *The Place* subtly unique.

THE CHOIRS OF DAY AND NIGHT

The primary sonic atmosphere of *The Place* is audible light. Just as the visible colors span the spectrum, a slow sweep of synthetic noise is filtered through two "harmonic prisms," banks of band-pass filters tuned to harmonies derived from the harmonic series and the subharmonic (inverted harmonic) series. As the noise passes through these prisms, two choirs of voice-like tones emerge, filling the space with slowly changing fields of audible colors.

Like the lighting, the noise sweep rises and falls continuously in relation to the rhythms of day and night and the height of the sun (H) above or below the horizon. When the sun is below the horizon, the pitch and amplitude of the noise are low. When the sun is above

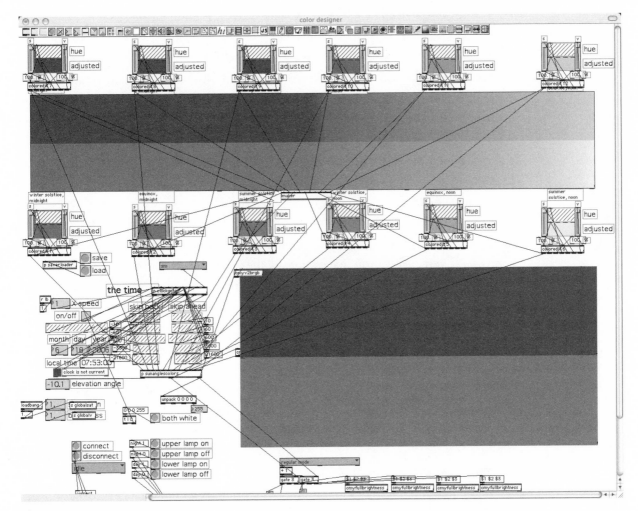

Figure 9.
Color program
for The Place.

the horizon, the pitch and amplitude rise. The audible center of the noise sweep also moves around the space according to the movements of the sun around the 360-degree horizon.

The sun sweep and all musical elements of *The Place* are tuned in relation to a fundamental frequency of 24.27 Hz. This fundamental is derived from the daily rotation of the earth, transposed into the range of human hearing. One rotation of the earth (one cycle per 86,400 seconds) transposed up twenty-one octaves yields a fundamental of 24.27 Hz.

Over the course of a year, at any given latitude and time of day, the height of the sun rises and falls approximately forty-seven degrees. In the musical world of *The Place,* forty-seven degrees produces a rise and fall of two octaves (2,400 cents) over the course of a year. This yields a rate of change of 51.0683 cents per degree of *H*.

In Fairbanks (64°51′30.3″ N, 147°50′38.4″ W) noontime *H* varies from about two degrees (on the winter solstice) to about forty-nine degrees (on the summer solstice). At solar noon on the vernal and autumnal equinoxes, the sun stands at about 25° *H*.

Table 3 lists approximate values for the height of the sun and the fundamental of the noise sweep in Fairbanks at solar noon on four important dates. Table 4 lists approximate values at solar midnight on the same dates.

TABLE 3.

2006	Height of Sun (Degrees)	Bottom Frequency (Hz)
Winter Solstice	2.0	107.93
Vernal Equinox	25.63	216.64
Summer Solstice	48.6	426.54
Autumnal Equinox	25.67	216.89

TABLE 4.

2006	Height of Sun (Degrees)	Bottom Frequency (Hz)
Winter Solstice	−48.58	24.27
Vernal Equinox	−24.48	49.42
Summer Solstice	−1.5	97.30
Autumnal Equinox	−24.83	48.91

Changes in cloud cover modulate the bandwidth of the noise sweep. When the sky is completely cloudless ("severe clear"), the bandwidth is four octaves (4,800 cents) wide. In partly cloudy conditions, it is three octaves (3,600 cents) wide. Under the cloudiest conditions, the bandwidth of the noise sweep diminishes to two octaves (2,400 cents) in width.

The continuous noise sweep is filtered through the harmonic prisms, one resonating with day, the other with night.

The Day Choir is grounded in the rising harmonic series, through the first eleven harmonics and their multiples. Harmonics 1 through 12, plus 14 and 15, are voiced as they occur in the first four octaves of the harmonic series. Higher octaves repeat the seven-tone sequence of the fourth octave. This yields bright, "major"-sounding harmonies.

The Night Choir is tuned as a subharmonic field, an inversion of the harmonic series. It also has an eleven-limit and octave doublings similar to the Day Choir. This produces dark, "minor"-sounding clouds of sound.

The Choirs of Day and Night are tuned to the tones shown in figure 10.

The Aurora Bells that sound with fluctuations in the magnetosphere are tuned to prime-number "just" intonations. The voice of the moon and the Earth Drums that sound with seismic activity employ continuous tones created with noise. But the primary harmonic fields of *The Place* are tuned in twelve-tone equal temperament.

Equal temperament was adopted only after repeated attempts to tune the Day Choir to just (whole-number) intervals. When the Day Choir was tuned to whole-number relationships, the harmonics reinforced one another, producing a strong buzzing in the fundamental tone. Equal temperament avoids this problem, producing beautiful sounds in all light and sky conditions throughout the year. Equal temperament accommodates acceptable approximations of whole-number tunings for all harmonic and subharmonic tones in the Night and Day Choirs. And its even distribution of pitches produces the desired effect of a diffuse sonic atmosphere.

Just as cloud cover modulates the bandwidth of the noise sweep, the visibility conditions of the sky over Fairbanks influence the bandwidth of the individual tones in the harmonic prisms. The tones in the Day Choir vary from 40 cents in conditions of "unlimited visibil-

Figure 10.
Tunings of the
Day Choir and
Night Choir.

Day Choir

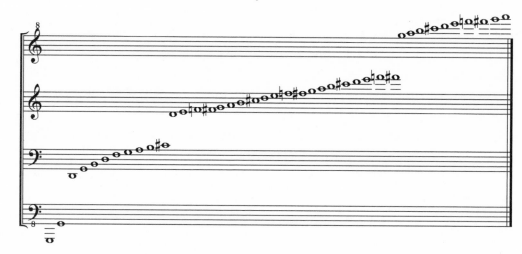

Night Choir

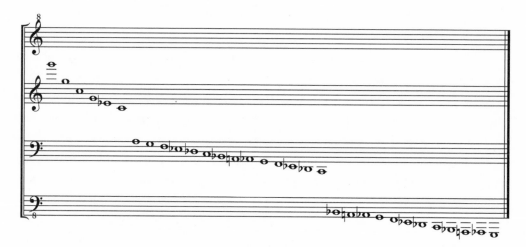

ity" to 60 cents in "zero visibility" conditions. The tones in the Night Choir vary from 20 cents in "unlimited visibility" to 60 cents in "zero visibility."

The combined sky conditions (both cloud cover and visibility) produce marked changes in the sounds of the Choirs. Under clear skies with unlimited visibility, the voices are bright, clear and sharply defined. Under cloudy skies with low visibility, the voices are darker, their timbres breathier and less distinct.

In addition to shaping the arcs of the noise sweep, the movements of the sun also control the spatial location and relative loudness of the Choirs. The Day Choir is always centered on the current location of the sun, while the Night Choir is always centered 180 degrees opposite.

Just as the fundamental pitch of the noise sweep is controlled by the height of the sun, so are the amplitudes of the harmonic fields.

The amplitude curves of the harmonic fields are plotted from five fixed points:

1. At solar noon on the summer solstice, the amplitude of the center of the Day Choir is −3 dB.
2. At solar midnight on the winter solstice, the amplitude of the center of the Night Choir is 0 dB.
3. At sunrise and sunset on any day (when the sun is at 0° H), the amplitude of the center of both Choirs is −6 dB.
4. At solar noon on the summer solstice, the amplitude of the center of the Night Choir is −12 dB.
5. At solar midnight on the winter solstice, the amplitude of the center of the Day Choir is −12 dB.

Between these points, the amplitudes of the Day and Night Choirs rise and fall with the height of the sun. Around sunrise and sunset, when both choirs are equally present, the tonal space of *The Place* is saturated with complex, chromatic harmonies.

To soften the sonic brightness of clear summer days, an amplitude "tilt" is applied to the upper octave of the Day Choir. From 0 dB at 1,701.5 Hz, the amplitude is rolled off to −3 dB at 3,415 Hz and on to −9 dB at 6,430 Hz.

A second curve determines the lateral "spread" of the Choirs from their central amplitudes. At any moment the difference between the center (loudest amplitude) of a harmonic field and the point 180 degrees opposite is −18 dB.

THE VOICES WITHIN

In their middle registers, the sounds of the Day and Night Choirs are especially evocative of human voices. This illusion is accentuated by a secondary sweep of noise nested within the larger noise sweep. This secondary sweep follows the same curves, with the same fundamental frequencies as the larger sweep. The bandwidth of the secondary sweep varies from two octaves (2,400 cents) under "severe clear" skies to one octave (1,200 cents) under the cloudiest sky conditions.

The secondary sweep passes through harmonic prisms with the same tunings as the Day and Night Choirs. As in the larger fields, the bandwidth of the individual tones varies with visibility. In this case, however, the bandwidth is much narrower—ranging from 1 cent in "unlimited visibility" to 20 cents in "zero visibility." When the sun is above the horizon (101.73 Hz), this evokes a smaller choir of solo voices that seem to undulate and breathe within the larger Choirs.

THE VOICE OF THE MOON

The moon also has a voice in the music of *The Place*.

Two interrelated programs, MoonAngles and MoonPhases, track the position and phase of the moon at any date and time. The voice of the moon rises and falls, circles the horizon, waxes and wanes with the cycles of the real moon in the real sky.

When the moon is below the horizon, it has a deep bass voice. As it rises, it glides continuously from baritone, to tenor, to alto, to soprano. The sound is produced by a narrow sweep of pink noise that follows the same pitch-to-height relationships as the Day and Night Choirs. The bandwidth of this sweep varies from about a Major Third (408 cents) at the full moon to silence (0 cents), briefly, at the new moon.

Figure 11.
The moon over Fairbanks, December 2006.
(Geophysical Institute of the University of Alaska.)

To create the perception of equal loudness when the moon is above the horizon, an amplitude curve is applied to its noise sweep. When the moon is at its zenith (53.6° H), its amplitude is 0 dB. When it is at the horizon (0° H), its amplitude is +6 dB. When it drops below the horizon, its amplitude drops off to 0 dB at its nadir (−53.6° H).

Table 5 shows a few examples of the sound of the moon at selected phases and heights in a single month (December 2006) (see also figure 11):

TABLE 5.

Date, Time	Phase	Height (Degrees)	Bottom Frequency (Hz)	Bandwidth (cents)	Amplitude (dB)
12/5, 00:59	1.0	53.06	486.5	408	1.0
12/9, 12:51	0.75	5.63	119.9	305	1.9
12/12, 5:46	0.50	29.83	231	204	−1.4
12/15, 00:18	0.25	−24.83	48.9	102	1.5
12/20, 5:18	0.0	−35.11	36.1	0	1.3

Just as the sounds of the Day and Night Choirs move slowly up and down and around the listening space, the sound of the moon also moves around *The Place* in conjunction with the movements of the real moon in the real sky.

AURORA BELLS

Just as the cycles of night and day create the primary sonic atmosphere of *The Place,* fluctuations in the magnetic field of the earth are also translated into music. When the aurora borealis is active, the listener hears shimmering veils of sound floating across the ceiling. These sounds are produced by arrays of virtual "bells."

The music of the Aurora Bells originates with geomagnetic activity measured by magnetometers in five locations from the Arctic coast to the south slope of the Alaska Range.

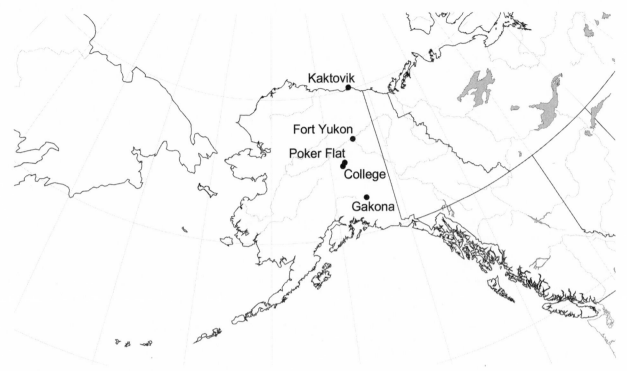

Figure 12.
Magnetometer stations of The Place.
(University of Alaska Museum of the North.)

Real-time data streams provided by the Geophysical Institute at the University of Alaska Fairbanks are translated into a sweep of pink noise that varies with the movement of the magnetic field over each location.

Major perturbations in the magnetosphere are closely associated with the aurora borealis. When the magnetosphere is active, so is the aurora. Yet the relationships between the spectral dance of the northern lights and the invisible fluctuations of geomagnetism are complex, making it virtually impossible to correlate the two with any degree of specific detail.

Although the Aurora Bells of *The Place* are "played" by the movements of the magnetic field, the sound forms they produce are not sonic illustrations or simulations of the aurora. Even so, when the aurora is active, the Aurora Bells are sounding. And because the bells are controlled by geomagnetism rather than the visible forms of the aurora, the listener in *The Place* can (metaphorically speaking) "hear the aurora" even on overcast nights or during daylight hours when it is not visible.

The noise sweeps that create the Aurora Bells are tuned relative to the sonic horizon (101.73 Hz) and the geographic location of *The Place,* distributed to fill the aural sky both harmonically and spatially.

The highest possible frequency in the noise sweep of the sun is 6,825 Hz. (This frequency would sound at solar noon on the summer solstice with "severe clear" sky conditions.) This is also the highest sounding frequency for the geomagnetic noise sweeps. The sweep associated with the College magnetometer will touch 6,825 Hz when geomagnetic activity over that station is at its maximum.

The full range of the geomagnetic noise sweeps of all five magnetometer stations is mapped from this zenith (90° H) to the sonic horizon (0° H) of *The Place* (101.73 Hz). This mapping yields a pitch-to-height relationship of 80.9 cents per degree H for the electromagnetic noise sweeps.

The upper and lower frequency of each station's noise sweep is defined by the upper and lower boundaries of a typical aurora borealis, as seen from *The Place.* This "aural perspective" is a function of the distance and direction of each station relative to *The Place.* When

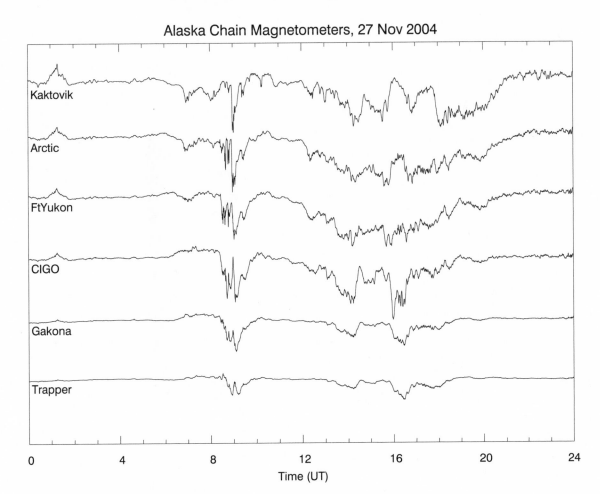

Figure 13.

Graph of magnetometer data.

(Geophysical Institute of the University of Alaska.)

TABLE 6.

Kaktovik (70°8′6″ N, 143°39′00″ W) (606 kilometers from *The Place*)

Upper Edge	23.63° *H*	306.65 Hz	1,910 cents (ç) above Horizon
Lower Edge	9.37° *H*	157.56 Hz	757 ç above Horizon

Fort Yukon (66°33′36″ N, 145°13′12″ W) (222 kilometers from *The Place*)

Upper Edge	53.15° *H*	1,216.71 Hz	4,296 ç above Horizon
Lower Edge	24.25° *H*	315.62 Hz	1,960 ç above Horizon

Poker Flat (65°7′8.4″ N, 147°25′48″ W) (56 kilometers from *The Place*)

Upper Edge	85.11° *H*	5,412.03 Hz	6,880 ç above Horizon
Lower Edge	60.75° *H*	1,735.15 Hz	4,910 ç above Horizon

College (64°52′22.8″ N, 147°51′36″ W) (1.8 kilometers from *The Place*)

Lower Edge	90° *H*	6,798.9 Hz	7,274 ç above Horizon
Upper Edge	82.11° *H*	4,704.5 Hz	6,637 ç above Horizon

Gakona (62°23′34.8″ N, 145°7′48″ W) (304 kilometers from *The Place*)

Upper Edge	36.44° *H*	557.58 Hz	2,945 ç above Horizon
Lower Edge	18.21° *H*	238.05 Hz	1,471 ç above Horizon

Note: The College magnetometer station is located very close to *The Place*. When aurora is visible directly overhead, the upper edge of the form appears lower than the upper edge. This is reflected in the sound mapping of magnetometer data in *The Place*.

geomagnetic activity is centered above a given station, the noise sweep encompasses the frequency ranges shown in table 6.

The magnetometers measure fluctuations in the magnetosphere in three directions:

H—measures the fluctuation of the magnetic field in the north–south plane
D—measures the fluctuation of the magnetic field in the east–west plane
Z—measures the fluctuation of the magnetic field in the vertical plane

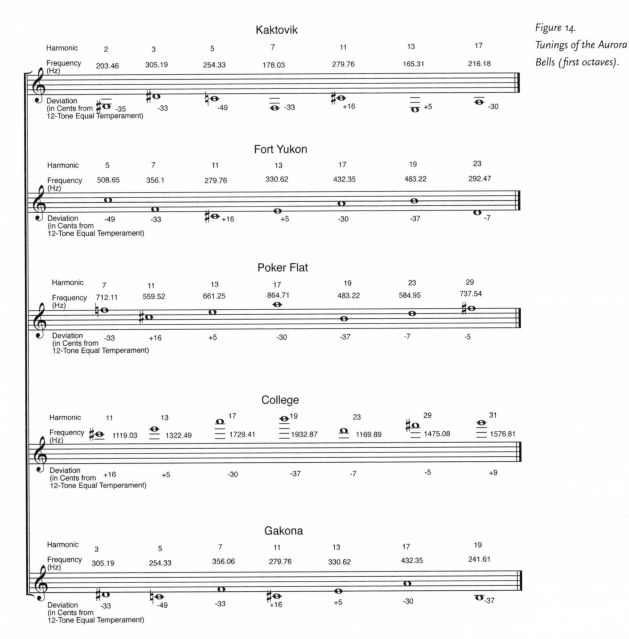

Figure 14.
Tunings of the Aurora
Bells (first octaves).

The combination of these three components creates a vector that describes the total fluctuation of the magnetic field. This vector drives the algorithm that generates the noise sweep associated with each of the five stations.

The magnetometer data are filtered to produce a continuous stream of numbers showing difference (in positive and negative values) from a running average. Generally speaking, the greater the difference from the running average, the higher the frequency, the greater will be the amplitude and the narrower the bandwidth of the noise sweep.

Just as the noise sweeps associated with night and day are filtered through harmonic prisms, the noise sweep for each of the magnetometer stations is filtered through a harmonic prism, tuned to one of five different series of prime numbers.

The relative amplitude levels of the bell stations are a function of distance, ranging from 0 dB at *The Place* to −6 dB at Kaktovik, as shown in table 7.

TABLE 7.

Station	Relative Amplitude (dB)
Kaktovik	−6
Fort Yukon	−2.25
Poker Flat	−0.57
College	0
Gakona	−2.99

EARTH DRUMS

In *The Place Where You Go to Listen* a set of virtual "drums" is sounded by seismic data provided by the Alaska Earthquake Information Center from five stations throughout Interior Alaska. During moderate to large earthquakes, low-frequency sounds rumble and boom through the space as different stations receive the seismic vibrations at different times and intensities.

The seismic stations heard in *The Place* are the following:

Figure 15.
Seismic stations of The Place.
(University of Alaska Museum of the North.)

Coldfoot (67°13′48″ N, 150°12′00″ W) (284.63 kilometers from *The Place*)

College (64°52′12″ N, 147°51′00″ W) (1.34 kilometers from *The Place*)

Paxson (62°58′12″ N, 145°28′12″ W) (240.15 kilometers from *The Place*)

Kantishna (63°30′00″ N, 150°55′12″ W) (212.34 kilometers from *The Place*)

Purkeypile (62°54′00″ N, 152°11′24″ W) (304.65 kilometers from *The Place*)

The majority of the natural cycles and events heard in *The Place* lie beyond the normal reach of human hearing. This is true for most of the seismic activity as well.

Most of the seismic events heard in the work are too low in energy or too distant to cross the threshold of feeling. However, since large earthquakes may occasionally be felt directly through the floor of the room, the seismic data streams are translated into sound in as close to real time as possible so that as soon as a visitor feels an earthquake in *The Place* she will hear it.

Even so, much of the time when the Earth Drums are sounding they are announcing seismic events that elude our unassisted perception. Alaska is one of the most seismically active places on earth. Approximately twelve thousand measurable earthquakes occur in the state each year. Most of these quakes are too low in magnitude to be felt through the ground, but many of them will be heard in *The Place*. The sensitivity of the Earth Drums is set to maximize the amount of seismic activity heard.

Like the drums, earthquakes produce highly dynamic waves. The range of energy they produce is extremely wide. In order to encompass such extremes, the dynamic range of the Earth Drums in *The Place* is compressed, allowing ground waves and small seismic events to be heard, while not overdriving the sound system during a major earthquake.

The seismometers measure movements of the earth in three components:

N—measures acceleration in a north–south direction

E—measures acceleration in an east–west direction

Z—measures acceleration in a vertical direction

An earthquake generates two types of waves: P waves (primary waves) and S waves (shear or secondary waves). Like sound waves, P waves are compression waves, except the medium

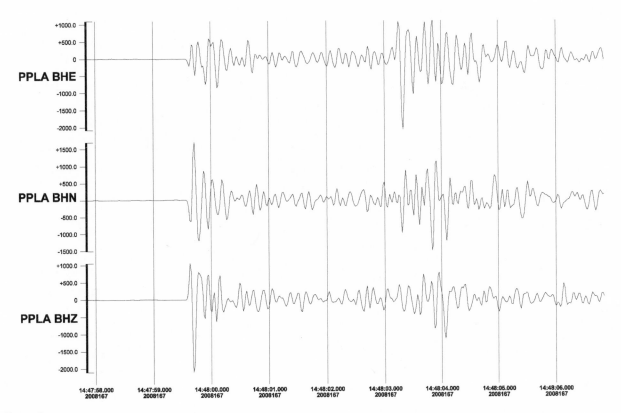

Figure 16.
Graph of seismic data.
(Geophysical Institute of the University of Alaska.)

through which they travel is not air, but rather the earth. P waves move along the wave path between the earthquake and the receiving station. S waves move orthogonally, perpendicular to the wave path.

Quite often the angle of incidence between the wave path and the surface of the ground is nearly vertical. This means that much of the P wave energy is observed in the vertical (Z) component, while the horizontal components (N and E) better record the S waves.

In the sound mapping of *The Place*, the N and E components are combined with white noise (through spectral convolution) to produce the drum sounds. The Z component is used to modulate the pitches of the drum sounds as seismic events occur. The greater the vertical seismic movement, the greater are the pitch fluctuations of the drums.

The range of pitch fluctuations varies with the distance of the seismic stations, from one octave (1,200 cents) at *The Place* to no fluctuation (0 cents) at 350 kilometers.

The center frequency of all the Earth Drums is 34.28 Hz. As seismic activity occurs, the pitches of the drums fluctuate above and below this frequency. During the largest, nearest seismic event, the lowest sound produced by the drums is 24.27 Hz—the fundamental frequency (derived from the earth's daily rotation) on which the entire sonic world of *The Place* is grounded. The highest sound produced by the drums is 48.42 Hz, well below the sonic horizon of 101.73 Hz—the metaphorical "ground level" of *The Place*.

Like the frequency ranges, the relative amplitude levels of the drum stations are also a function of distance, ranging from 0 dB at *The Place* to −6 dB at 350 kilometers.

Time delay adds another layer of perceived depth to the sounds of the drums. Since the seismic stations are many miles apart, a given seismic event will be recorded at a different moment at each station. These differences produce a series of echoes in the Earth Drums. The larger the seismic event, the louder and longer the echoes of the drums will be.

Since the Earth Drums produce very low frequency sounds, they are heard only through the subwoofer speakers hidden in the east and west walls of *The Place*. This means the sounds of the drums cannot be localized in space the way the sounds of day and night, the moon and the Aurora Bells are. The drums are localized only on the west–east axis (from 0.00 for due west to 1.00 for due east).

Table 8 shows the frequency ranges, relative amplitude levels and localization for data from each of the seismic stations, as heard in *The Place*.

TABLE 8.

Station	Frequency Range (cents)	Relative Amplitude (dB)	Localization
College	1,196	0	0.65
Kantishna	474	−3.63	0.19
Paxson	378	−4.11	1.00
Coldfoot	226	−4.87	0.30
Purkeypile	156	−5.2	0.00

THE SONIC SPACE

Sound in *The Place* is distributed throughout the space according to the movements of the sun and the moon, as well as the geographic locations of the magnetometers and seismometer stations.

The sound system includes twelve main speakers, three each in the south wall, north wall, south ceiling and north ceiling. In the vertical plane, the wall speakers are centered below the nominal horizon of *The Place,* at −30° *H.* The ceiling speakers are centered above the horizon, at +60° *H.* In the horizontal plane, the wall speakers are centered at the following locations on the 360° azimuth horizon.

Southeast	120°
South Central	180°
Southwest	240°
Northeast	60°
North Central	0°/360°
Northwest	300°

The subwoofer speakers are centered at 90 degrees (due east) and 270 degrees (due west).

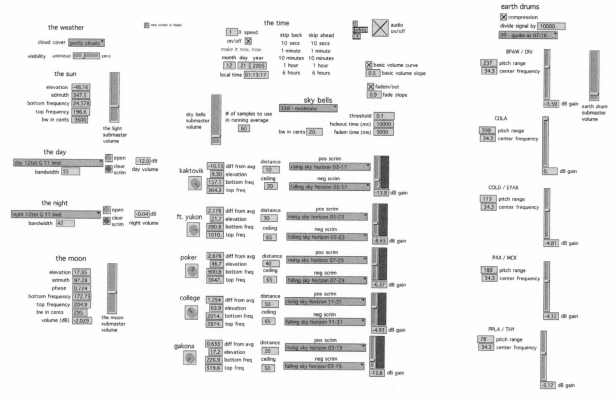

Figure 17.
Main program for The Place.

The sounds of the Aurora Bells are centered at these locations above the northern horizon.

Kaktovik	$24°\ H$
Fort Yukon	$53°\ H$
Poker Flat	$85°\ H$
College	$90°\ H$
Gakona	Centered $36°\ H$ above the southern horizon

THE ONE WHO LISTENS

She listened. And she heard.

But she rarely spoke of these things.

She did not question them.

This is the way it is for one who listens.

The Place Where You Go to Listen was conceived of as a self-contained sound world in which the inaudible becomes audible. From the outset I knew that no familiar musical instruments or recognizable sounds from nature would be heard in this world. I was determined to avoid conventional modes of musical expression and picturesque evocations of nature.

At the same time I wanted *The Place* to be directly connected to nature, and I wanted these connections to have the palpable resonance of the real. Even if the listener knows little about the natural forces behind the work, I wanted *The Place* to sound and feel *natural.* And it's my hope that this work may inspire the listener to new perceptions of the larger world in which we live.

The world contains both environments and ecosystems. An environment is a natural system of elements and forces interacting in a place. An ecosystem is the totality of an environment and the organisms that inhabit it functioning together as a whole.

When no listener is present in *The Place Where You Go to Listen,* the forces of nature continue to sound within the environment of the work. But when a listener is present, her awareness transforms this sound environment into an ecosystem of music.

Entering the antechamber of *The Place,* the visitor encounters the following text:

THE PLACE WHERE YOU GO TO LISTEN

We are immersed in music.

The earth beneath us, the air around us, and the sky above us are filled with vibrations. Most of these vibrations are beyond the reach of our ears.

In this room you will hear some of this music.

You will hear no familiar musical instruments or sounds of nature. Yet every sound you hear is connected directly to the natural world, here and now.

The atmosphere of sound and light changes with the movements of the sun, the rhythms of day and night. Daylight sings like a choir of bright voices. Its colors are yellow, orange and red. The voices of night are darker. Its colors are violet, blue and cyan.

The moon rises and falls, appears and disappears, like a solo voice.

When the aurora borealis is active (even if hidden by daylight or clouds) bell-like sounds float across the ceiling.

When the earth quakes (even imperceptibly) the walls and the floor shudder and rumble like deep drums.

This music has no beginning, middle or end. Even in moments of apparent stillness, it is always changing. But it changes at the tempo of nature. To experience its full range requires listening in day and night, winter and summer.

This is an ecosystem of sound and light that resonates with the larger world around it. When no one is here, the forces of nature continue to reverberate within this space.

But the awareness of the listener brings it to life.

The Place Where You Go to Listen is not complete until you are present and listening.

Afterword

From the beginning I hoped *The Place Where You Go to Listen* would touch some of the thousands of people from all over the world who visit the Museum of the North each year. Perhaps even more I hoped it might have meaning for people who live here in Alaska. This is, after all, my own home.

Since it first opened to the public, I've heard from many people who've told me about their experiences inside *The Place* and what it means to them. Every story is different, and I listen to each one with curiosity and delight. I'm amazed by some of the experiences people describe and the insights they share with me. They often hear things I haven't heard before. And they understand *The Place* in ways I hadn't understood it myself.

By now there are a number of people who have spent more time inside *The Place* than I have. Some who work at the museum visit *The Place* almost every day. Others who work nearby visit once or twice a week. Some people meditate inside *The Place*. Others write or sketch. Still others just sit and listen. Schoolteachers bring their classes in and ask them later to draw or write about their experiences. A number of local residents have told me they like to bring their out-of-town guests to *The Place*. The many ways that people have made *The Place* their own have far exceeded my hopes.

Before it opened, I wasn't sure how visitors might react to *The Place*. This is not extroverted work. It doesn't seek to entertain. And it doesn't reveal its secrets easily. *The Place* requires the listener to take time, to actively participate in shaping her own experience. The casual visitor may walk in, look at the colors, listen for a few seconds and walk out, saying: "I don't get it." Yet I've been delightfully surprised by how many people really *do* "get it."

One of the most challenging aspects of *The Place* is its time scale. When a seismic event or an electromagnetic storm is under way, the room is filled with dynamic sounds. But more often than not, there's little drama and no fireworks. Even so, the majority of visitors seem to understand the "real-time" nature of *The Place*—that each visit is a unique experience of the colors, the ambience, the *feeling* of each particular moment. And they seem to understand that this is an essential part of the experience.

When I began *The Place,* I imagined that it could be realized anywhere on the earth, tuned to any geographic location. As the work evolved, I came to regard it as unique to this particular geography. Now, several years later, I see it both ways.

The Place grew out of my abiding love for Alaska. This work might never have happened, and it certainly wouldn't be the same in any other place. Yet I now imagine this as the first of a series of listening stations to be created in sites around the earth. And in time *The Place Where You Go to Listen* may prove to be a longer, more open-ended journey than I yet understand.

Acknowledgments

Jim Altieri – Programmer

James Brader – Meteorology

Roger Hansen – Seismology

Dirk Lummerzheim – Aurora

John Olson – Physics and Infrasound

Josh Stachnik – Seismology programmer

Curt Szuberla – Mathematics, Programmer of SunAngles, MoonAngles, MoonPhases

Roger Topp – Interactive audio–visual program

Debi-Lee Wilkinson – Magnetometer data and programming

John Luther Adams photo by Dennis Keeley

From his home in Alaska, John Luther Adams has created a unique musical world grounded in wilderness landscapes and natural phenomena from the songs of birds to elemental noise. His music includes works for orchestra, small ensembles, percussion and electronic media, recorded on Cold Blue, New World, Cantaloupe, Mode, and New Albion.

His sound and light environment *The Place Where You Go to Listen* is a long-term exhibition at the Museum of the North at the University of Alaska Fairbanks. He is the author of *Winter Music* (Wesleyan 2004), and his writings have appeared in numerous periodicals and anthologies.

Born in 1953, Adams grew up in the South and in the suburbs of New York City. He studied composition with James Tenney at the California Institute of the Arts, where he was in the first graduating class (BFA 1973). In the mid 1970s he became active in the campaign for the Alaska National Interest Lands Conservation Act, and subsequently served as executive director of the Northern Alaska Environmental Center.

Adams has worked with many prominent performers and venues, including Bang On A Can, the EAR Unit, Percussion Group Cincinnati, the Paul Dresher Ensemble, Other Minds, Radio Nederlands, Almeida Opera and the Sundance Institute. He has been composer in residence with the Anchorage Symphony, Anchorage Opera, Fairbanks Symphony, Arctic Chamber Orchestra and the Alaska Public Radio Network, and served as president of the American Music Center. He has taught at the University of Alaska, Bennington College and the Oberlin Conservatory of Music.

In 2006 Adams was named one of the first United States Artists Fellows. He has also received fellowships from the Rasmuson Foundation, the Foundation for Contemporary Arts and the National Endowment for the Arts.

Currently (2009) Adams is at work on a concert piece for large percussion ensemble to be premiered outdoors in the Canadian Rockies at the Banff Centre, and planning a new sound-and-light environment for the University of California San Diego.

Bibliography and Discography

COMPILED BY NOAH POLLACZEK

The works of John Luther Adams, composed over a period of more than thirty years, have revealed a singular and distinctive musical voice. Through recordings, publications and public performances, his contributions to contemporary music have been considerable. This bibliography aims to inventory Adams's complete oeuvre in three parts, through an enumeration of his works, recordings and writings. Also included is an inventory of selected writings about the composer, which illustrate his influence on the broader musical culture.

Numerous resources proved useful in the compilation of this bibliography. Sabine Feisst's discography and catalog of works for Adams's *Winter Music: Composing the North* served as valuable points of reference and frequently provided key information regarding first performances of Adams's compositions. The Web site of John Luther Adams (www.johnlutheradams.com) offered additional information regarding sound recordings, compositions, musical performances and writings.

SELECTED WORKS

LISTED ALPHABETICALLY

Always Very Soft (1973/2007)
for three percussionists   9:00
Dedicated to Percussion Group Cincinnati

Among Red Mountains (2001)
for piano   10:30
Dedicated to Peter Garland
First performed December 7, 2001, by Emily Manzo at Oberlin College, Oberlin, Ohio
Recorded on *Red Arc/Blue Veil* (Cold Blue CB0026)

. . . and bells remembered . . . (2005)
for crotales (bowed), orchestra bells, chimes, vibraphone (bowed), vibraphone (struck)   10:15
Commissioned by the University of Wisconsin–River Falls
First performed 2006 at the University of Wisconsin–River Falls

Clouds of Forgetting, Clouds of Unknowing (1991–95)
for two violins, viola, 'cello, doublebass, two flutes/piccolos, clarinet, bass clarinet, two horns,
    trumpet, bass trombone, celesta, piano, two percussionists   62:00
First performed 1996 by JoAnn Falletta and the ensemble Apollo, Norfolk, Virginia
Recorded on *Clouds of Forgetting, Clouds of Unknowing* (New World 80500-2)

Crow and Weasel (1993–94)
for string quintet, piccolo/bass clarinet, celesta, harp, four percussionists   90:00
Music written for the theatrical production; story by Barry Lopez, stage adaptation by
    Jim Leonard, script published by Samuel French
Commissioned by the Sundance Institute and the Children's Theater Company
First performed 1994 by John Luther Adams and Musicians, Minneapolis, Minnesota

Dark Waves (2007)
for orchestra and electronically processed sounds   12:00
Commissioned by Music Nova, for the Anchorage Symphony Orchestra
First performed 2007 by Randall Craig Fleischer and the Anchorage Symphony, Anchorage, Alaska
Recorded by Radio Nederlands Filharmonish; conducted by Jaap van Zweden (2007)

Dark Waves (2007)
version for two pianos and electronically processed sounds   12:00
First performed 2007 by Stephen Drury and Yukiko Takagi, Boston, Massachusetts
Recorded on *Red Arc/Blue Veil* (Cold Blue CB0026)

Dark Wind (2001)
for bass clarinet, piano, vibraphone, marimba   13:15
Commissioned by Marty Walker
First performed 2002 at Vanderbilt University, Nashville, Tennessee
Recorded on *Adams/Cox/Fink/Fox* (Cold Blue CB0009)

Dream In White On White (1992)
for string quartet, harp (or piano), strings   16:45
First performed May 25, 1992, by JoAnn Falletta and the ensemble Apollo at Old Dominion
    University, Norfolk, Virginia
Recorded on *The Far Country* (New Albion NA061)

Earth and the Great Weather (A Sonic Geography of the Arctic) (1990–93)
for violin, viola, 'cello, doublebass, two sopranos, alto, bass, four speaking voices, four
    percussion, digital delay, recorded natural sounds   90:00
Commissioned by the Alaska Festival of Native Arts
First performed 1993 by John Luther Adams and Company at the University of Alaska Fairbanks
Recorded on *Earth and the Great Weather* (New World 80459-2)

The Far Country of Sleep (In Memoriam Morton Feldman) (1988)
for orchestra   15:45
Commissioned by the Arctic Chamber Orchestra
First performed 1988 by John Luther Adams and the Arctic Chamber Orchestra, Haines, Alaska
Recorded on *The Far Country* (New Albion NA061)

The Farthest Place (2001)
for violin, doublebass, piano, vibraphone, marimba   11:00
Recorded on *The Light That Fills the World* (Cold Blue CB0010)

Five Athabascan Dances (1992/1996)
for harp and percussion   16:00
Commissioned by the U.S. Embassy, Tokyo, Japan
First performed by Just Strings, Tokyo, Japan

Five Yup'ik Dances (1991–94)
for harp   12:00
First performed 1991 by Heidi Lehwalder, Seattle, Washington

For Jim (Rising) (2006) (In Memory of James Tenney)
for three trumpets and three trombones   6:30
First performed October 30, 2007, by the Orkest de Volharding, Rotterdam, Netherlands

for Lou Harrison (2003)

for string quartet, two pianos, strings   65:00

First performed September 27, 2005, by Stephen Drury and the Callithumpian Consort at the
New England Conservatory of Music, Boston, Massachusetts

Recorded on *for Lou Harrison* (New World 80669-2)

Giving Birth to Thunder, Sleeping With His Daughter, Coyote Builds North America (1986/90)

Music written for the theatrical production; text by Barry Lopez

for violin, double bass, e-flat clarinet, bass clarinet, four percussionists, storyteller   70:00

Commissioned by Perseverance Theater

First performed 1987 by John Luther Adams and Musicians, with Perseverance Theater,
Juneau, Alaska

Green Corn Dance (1974)

for six percussionists   7:30

Dedicated to James Tenney

First performed March 26, 1977, by the New Music Ensemble at Memphis State University,
Memphis, Tennessee

The Immeasurable Space of Tones (1998/2001)

for violin, contrabass instrument (doublebass, contrabass clarinet, or contrabassoon), piano,
vibraphone, electronic keyboard (or organ)   27:00

Recorded on *The Light That Fills the World* (Cold Blue CB0010)

In the White Silence (1998)

for string quartet, celesta, harp, two vibraphones, strings   75:00

First performed November 11, 1998, by Tim Weiss and the Oberlin Contemporary Music
Ensemble in Finney Chapel at Oberlin College, Oberlin, Ohio

Recorded on *In the White Silence* (New World 80600-2)

The Light That Fills the World (1999–2000)

for violin, contrabass instrument (doublebass, contrabass clarinet, or contrabassoon),
vibraphone, marimba, electronic keyboard (or organ)   13:00

Commissioned by the Paul Dresher Ensemble

First performed 1999 by the Paul Dresher Ensemble, San Francisco, California
Recorded on *The Light That Fills the World* (Cold Blue CB0010)

The Light That Fills the World (1999–2000)
version for orchestra   13:00
First performed 2000 by Gordon Wright and the Fairbanks Symphony Orchestra, Healy, Alaska

The Light Within (2007)
for violin, 'cello, alto flute, bass clarinet, piano, vibraphone, crotales, electronically
    processed sounds   12:00
Commissioned by the Seattle Chamber Players and the California EAR Unit
First performed January 28, 2008, by the Seattle Chamber Players, Seattle, Washington

Make Prayers to the Raven (1996–98)
for violin, 'cello, flute, harp (or piano), percussion   16:30
First performed 2000 by the Min Ensemble, Bergen, Norway

The Mathematics of Resonant Bodies (2002)
for solo percussion, electronically processed sounds   70:00
Commissioned by the Los Angeles County Museum of Art, WNYC and the Subtropics Festival.
First performed April 28, 2003, by Steven Schick at the Los Angeles County Museum of Art,
    Los Angeles, California
Recorded on *The Mathematics of Resonant Bodies* (Cantaloupe CA 21034)

Night Peace (1976–77)
for antiphonal choirs, solo soprano, harp, percussion   14:00
Commissioned by the Atlanta Singers
First performed 1977 by Kevin Culver and the Atlanta Singers, Atlanta, Georgia
Recorded on *A Northern Suite/Night Peace* (Opus One 88) and *The Far Country*
(New Albion NA061)

A Northern Suite (1979–1980; revised 2004)
for orchestra   19:00
Commissioned by the Arctic Chamber Orchestra

First performed 1981 by Gordon Wright and the Arctic Chamber Orchestra, Galena, Alaska

Recorded on *A Northern Suite/Night Peace* (Opus One 88)

Nunataks (Solitary Peaks) (2007)

for piano   7:00

Commissioned by Music Northwest

First performed March 29, 2008, by Jane Harty, Seattle, Washington

The Place Where You Go to Listen (2002–6)

Sound and light environment, started March 20, 2006, at the University of Alaska Fairbanks,
    Museum of the North; ongoing

Qilyaun (1998)

for four bass drums (or bass drum with electronic delay)   15:00

Commissioned by the Fairbanks Symphony Association; dedicated to Scott Deal

First performed 1998 by Scott Deal, Fairbanks, Alaska

Recorded on *Red Arc/Blue Veil* (Cold Blue CB0026)

Red Arc/Blue Veil (2002)

for mallet percussion, piano, electronically processed sounds   12:00

Commissioned by Ensemble Sirius

First performed 2002 by Ensemble Sirius, Boston, Massachusetts

Recorded on the accompanying CD to *Winter Music* (Wesleyan University Press) and on
    Red Arc/Blue Veil (Cold Blue CB0026)

Sky With Four Suns/Sky With Four Moons (2007–8)

for four choirs   8:00

Commissioned by the Chamber Choir Kamer

First performed July 3, 2008, by Chamber Choir Kamer, Riga, Latvia

songbirdsongs (1974–79)

for two piccolos and three percussionists   ca: 40:00

First performed 1974 by John Luther Adams and Musicians, Atlanta, Georgia

Recorded on *songbirdsongs* (Opus One 66) and *Exotic Chamber Music* (Centaur CRC 2273)

Strange Birds Passing (In Memory of Tadashi Miyashita) (1983)
for two piccolos, three flutes, two alto flutes, bass flute   6:30
Dedicated to Dorli McWayne and the Fairbanks Flutists
First performed 1983 by the Fairbanks Flutists, Fairbanks, Alaska

Strange and Sacred Noise (1991–97)
for four percussionists   75:00
First performed 1998 by Percussion Group Cincinnati, Cincinnati, Ohio
Recorded on *Strange and Sacred Noise* (Mode 153)

Three High Places (In Memory of Gordon Wright) (2007)
for solo violin   10:00
First performed January 2008 by Erik Carlson, New York, New York

Veils and Vesper (2005)
Sound installation
First installed May 6–27, 2006, at the Diapason Gallery, New York, New York

SELECTED DISCOGRAPHY

LISTED CHRONOLOGICALLY

songbirdsongs   Opus One 66
1980   LP; 39:00
Anne McFarland and Michel Cook, piccolos and ocarinas; John Luther Adams, Kevin Culver,
Scott Douglas, Tim Embry, percussion and ocarinas

A Northern Suite/Night Peace   Opus One 88
1982   LP; 37:12
Arctic Chamber Orchestra; Gordon Wright, conductor
Atlanta Singers; Kevin Culver, conductor
Cheryl Bray, soprano; Joan Rubin, harp; Billy Traylor, percussion

Forest Without Leaves   Owl 32
1987   LP; 52:34
Arctic Chamber Choir and Arctic Chamber Orchestra; Byron McGilvray, conductor

The Far Country   New Albion NA061
1993   CD; 48:08
Dream In White On White
Apollo Quartet and Strings; JoAnn Falletta, conductor
Night Peace
Atlanta Singers; Kevin Culver, conductor
Cheryl Bray Lower, soprano; Nella Rigel, harp; Michael Cebulski, percussion
The Far Country of Sleep
Cabrillo Festival Orchestra; JoAnn Falletta, conductor

Earth and the Great Weather   New World 80459-2
1994   CD; 75:50
Robin Lorentz, violin; Ron Lawrence, viola; Michael Finckel, 'cello; Robert Black, doublebass;
John Luther Adams, Robert Black, Amy Knoles, Robin Lorentz, percussion; James Nageak,
Doreen Simmonds, Adeline Peter Raboff, Lincoln Tritt, Dave Hunsaker, spoken voices;
John Luther Adams, Michael Finckel, conductors

Clouds of Forgetting, Clouds of Unknowing   New World 80500-2
1997   CD; 61:24
Apollo; JoAnn Falletta, conductor

Adams/Cox/Fink/Fox   Cold Blue CB0009
2002   CD; 41:01
Dark Wind [13:22]
Marty Walker, bass clarinet; Bryan Pezzone, piano; Amy Knoles, vibraphone/marimba

The Light That Fills the World   Cold Blue CB0010
2002   CD; 50:58
Robin Lorentz, violin; Barry Newton, doublebass; Marty Walker, contrabass clarinet;
Amy Knoles, vibraphone/marimba; Nathaniel Reichman, keyboard/sound design

The Farthest Place
Robin Lorentz, violin; Barry Newton, doublebass; Bryan Pezzone, piano; Amy Knoles, vibraphone/marimba
The Immeasurable Space of Tones
Robin Lorentz, violin; Barry Newton, doublebass; Marty Walker, contrabass clarinet; Bryan Pezzone, piano; Amy Knoles, vibraphone; Nathaniel Reichman, keyboard/ sound design

In the White Silence   New World 80600-2
2003   CD; 75:07
Oberlin Contemporary Music Ensemble; Tim Weiss, conductor

Winter Music: Composing the North   Wesleyan University Press
2004   CD; 32:51
Selections included with Adams's book "Winter Music"
roar
Steven Schick, percussion
velocities crossing in phase-space
Percussion Group Cincinnati
Red Arc/Blue Veil
Ensemble Sirius

Strange and Sacred Noise   Mode 153
2005   DVD/CD; 73:06
Percussion Group Cincinnati
Russell Burge, James Culley, Allen Otte
with Stuart Gerber, Brady Harrison, Matt McClung
DVD includes the video interview "A Brief History of Noise: John Luther Adams and Allen Otte in Conversation"

The Mathematics of Resonant Bodies   Cantaloupe CA 21034
2006   CD; 69:09
Steven Schick, percussion

for Lou Harrison   New World 80669-2
2007   CD; 66:00
Callithumpian Consort; Stephen Drury, conductor

Red Arc/Blue Veil   Cold Blue CB0026
2007   CD; 51:48
Dark Waves
Stephen Drury, Yukiko Takagi, piano
Among Red Mountains
Stephen Drury, piano
Red Arc/Blue Veil
Stephen Drury, piano; Scott Deal, vibraphone/crotales
Qilyaun
Scott Deal, Stuart Gerber, bass drums

SELECTED WRITINGS BY JOHN LUTHER ADAMS

LISTED ALPHABETICALLY

Credo (In Memory of Gordon Wright)
Limited-edition broadside of ninety-seven signed and numbered copies; hand-set and printed by Mary Ellen Niedenfuer at Midnight Sun Paper Sales (June 2007); online at http://www.johnlutheradams.com/writings/credo.html.
"From the Ground Up"
Utne Reader, no. 68 (March–April 1995), p. 86
"Forest Without Leaves"
Ear Magazine, vol. 10, no. 3 (January–March 1986), pp. 5 and 20; abbreviated version in *Winter Music: Composing the North* (Middletown, Conn.: Wesleyan University Press, 2004), pp. 37–39

"Global Warming and Art"

    Musicworks, no. 86 (Summer 2003), pp. 8–9; reprinted in *Orion*, vol. 22, no. 5 (September–October 2003), p. 11; and *Winter Music: Composing the North* (Middletown, Conn.: Wesleyan University Press, 2004), pp. 177–83; originally published in slightly different form as "Alaska Is Melting: Can Art Help?" *Anchorage Daily News* (December 4, 2002)

"The Immeasurable Space of Tones"

    In *Winter Music: Composing the North* (Middletown, Conn.: Wesleyan University Press, 2004), pp. 162–64; abbreviated version in *Musicworks*, no. 91 (Spring 2005), pp. 7–8

"The Place Where You Go to Listen"

    Terra Nova, vol. 2, no. 3 (Summer 1997), pp. 15–16; reprinted in *North American Review*, vol. 283, no. 2 (March–April 1998), p. 35; and *The Book of Music & Nature: An Anthology of Sounds, Words, Thoughts* edited by David Rothenberg and Marta Ulvaeus (Middletown, Conn.: Wesleyan University Press, 2001), pp. 181–82; and online at http://www.johnlutheradams.com/writings/place.html

"Remembering James Tenney"

    Musicworks, no. 98 (Summer 2007), pp. 7–8

"Remembering Lou Harrison"

    New Music Box, May 7, 2003; available online at http://www.newmusicbox.org/article.nmbx?id=3560

"Resonance of Place: Confessions of an Out-of-Town Composer"

    North American Review, vol. 279, no. 1 (January–February 1994), pp. 8–18; available online at http://www.johnlutheradams.com/writings/resonance.html

"Sonic Geography: Alaska"

    Musicworks, no. 93 (Fall 2005), p. 5; originally published in slightly different form as "Aesthetic Belongs to Artist, Not to North," *Anchorage Daily News* (September 11, 2005)

"Strange and Sacred Noise"

    In *Northern Soundscapes*, Yearbook of Soundscape Studies, vol. 1, edited by R. Murray Schafer and Helmi Järviluoma (Tampere, Finland: University of Tampere, Dept. of Folk Tradition, 1998), pp. 143–46

"Visitations"

    Manoa, vol. 20, no. 1 (June 2008)

Winter Music: Composing the North
(Middletown, Conn.: Wesleyan University Press, 2004, pp. 57–77); reprinted as "Winter
Music: A Composer's Journal," *Musicworks,* no. 82 (Winter 2002), pp. 32–39; also *Reflections
on American Music* (Hillsdale, N.Y.: Pendragon Press, 2000), pp. 31–48; and in *The Best
Spiritual Writing 2002,* edited by Philip Zaleski (New York: HarperCollins Publishers),
pp. 1–21

SELECTED INTERVIEWS AND WRITINGS ABOUT JOHN LUTHER ADAMS

LISTED CHRONOLOGICALLY

Bernd Herzogenrath, "The Weather of Music: Sounding Nature in the 20th and 21st Century,"
Deleuze/Guattari & Ecology, edited by Bernd Herzogenrath (Basingstoke: Palgrave
Macmillan, 2008), pp. 216–232

Alex Ross, "Song of the Earth," *The New Yorker* (May 12, 2008), pp. 76–81

Peter Garland, *for Lou Harrison* (New World 80669-2) (2007) (liner notes), pp. 2–12; available
online at http://www.newworldrecords.org/uploads/file8Wiaq.pdf

Steven Schick, section on *The Mathematics of Resonant Bodies* in *The Percussionist's Art: Same
Bed, Different Dreams* (Rochester, N.Y.: University of Rochester Press, 2006), pp. 79–89

Robert Carl, "The Mathematics of Resonant Bodies," *Fanfare* vol. 30, no. 2 (November–
December, 2006), pp. 105–106

Robert Carl, "Strange and Sacred Noise," *Fanfare* vol. 29, no. 6 (July–August, 2006), pp. 40–41

Molly Sheridan, "Cold Spell," *Time Out New York* (May 4–10, 2006)

Amy Mayer, "Northern Exposure," *Boston Globe* (April 16, 2006), M1

Kyle Gann, "Long Ride in a Slow Machine," *NewMusicBox* (March 29, 2006); available online at
http://www.newmusicbox.org/article.nmbx?id=4573

Mike Dunham, "The Harmony of the Spheres," *Orion* (March–April, 2006)

Alan Gimbel, "Winter Music," *American Record Guide* vol. 69, no. 1 (January–February 2006),
pp. 295–297

Steven Schick, *Strange and Sacred Noise* (Mode 153) (2005), liner notes

Denise Von Glahn, "Winter Music," *Notes* vol. 62, no. 2 (December 2005), pp. 378–379

Rupert Loydell, "The Great White Wonder," *Tangents* (November 2004)

Molly Sheridan, "In Conversation with John Luther Adams," *NewMusicBox* (October 25, 2004); available online at http://www.newmusicbox.org/article.nmbx?id=3296

Amy Mayer, "Meditative Music from Alaska," *All Things Considered* (September 26, 2004); available online at http://www.npr.org/templates/story/story.php?storyId=3937518

Sabine Feisst, *In the White Silence* (New World 80600-2) (2003), liner notes; available online at http://www.newworldrecords.org/liner_notes/80600.pdf

Alan Gimbel, "In the White Silence," *American Record Guide* vol. 66, no. 5 (September–October, 2003), p. 66

Kyle Gann, "Erasing the Lines: John Luther Adams Explores a New Landscape of Pure Harmony," *Village Voice* vol. 48, no. 25 (June 18–24, 2003), p. 101

Kyle Gann, "American Composer: John Luther Adams," *Chamber Music* vol. 19, no. 1 (February 2002), pp. 46–47

Baker's Biographical Dictionary of Musicians, 8th ed., vol. 1, edited by Nicolas Slonimsky and Laura Kuhn (New York: Schirmer, 2001), pp. 16–17

Sabine Feisst, "Klanggeographie–Klanggeometrie. Der US-amerikanische Komponist John Luther Adams," *Musiktexte* vol. 91 (November 2001), pp. 4–14

Mark Alburger, "The Big Picture," *21st Century Music* vol. 8, no. 10 (October 2001), p. 19

Kyle Gann in H. Wiley Hitchcock, *Music in the United States* (Upper Saddle River, N.J.: Prentice-Hall, 2000), pp. 381–382

Joshua Kosman, *The New Grove Dictionary of Music and Musicians,* 2nd ed., vol. 1, edited by Stanley Sadie and John Tyrell (New York: MacMillan, 2000), p. 146

Phil England, "A Music Drama Celebrating Alaska's Arctic Refuge," *The Wire* 199 (September 2000), p. 66

David Bündler, "Recent Works by John Luther Adams," *21st Century Music* vol. 7, no. 2 (February 2000), p. 14

Mark Alburger, "A to Z: Interview with John Luther Adams," *21st Century Music* vol. 7, no. 1 (January 2000), pp. 1–12

Toshie Kakinuma, *Avant Music Guide* (Tokyo: Sakuhinsha, 1999), pp. 238–239

Mitchell Morris, "Ecotopian Sound or the Music of John Luther Adams and Strong Environmentalism," *Crosscurrents and Counterpoints: Offerings in Honor of Bengt Hambraeus at 70*, edited by P. F. Broman, N.A. Engebretsen and B. Alphonce (Göthenburg: University of Sweden Press, 1998), pp. 129–141

Gayle Young, "Sonic Geography of the Arctic. An Interview with John Luther Adams," *Musicworks* vol. 70 (Spring 1998), pp. 38–43

Kyle Gann, "A Forest from the Seeds of Minimalism. An Essay on Postminimal and Totalist Music," Program Notes for the Minimalism Festival of the Berliner Gesellschaft für Neue Musik (1998)

Kyle Gann, *American Music in the Twentieth Century* (New York: Schirmer, 1997), pp. 368–371

Howard Klein, *Clouds of Forgetting, Clouds of Unknowing* (New World 80500-2) (1997), liner notes; available online at www.newworldrecords.org/linernotes/80500.pdf

Mike Dunham, "In Review: From Around the World," *Opera News* vol. 60, no. 3 (September 1995), p. 62

Liane Hansen, "Alaska Inspires New Soundscape from John Luther Adams," *Weekend Edition, National Public Radio* (February 12, 1995)

Howard Klein, *Earth and the Great Weather* (New World 80459-2) (1994), liner notes; available online at www.newworldrecords.org/linernotes/80459.pdf

Thomas B. Harrison, "To Hear the Unheard," *Alaska Magazine* (February 1993), p. 96

Eric Salzman, "Two John Adamses," *Stereo Review* vol. 46, no. 10 (1981), p. 140